W9-BFC-249

General Key Characters		
Habit	Petal Number	Leaf Arrangement
T tree	N none, or not evident	N none, or not evident
S shrub	3	O opposite
V vine	4	A alternate
H terrestrial herb	5	W whorled
A aquatic herb	6	B basal
E epiphyte, parasite	V 7+ or variable	

Key Leaf Characters	
Leaf Type	Leaf Shape
S simple	G needle, scale, or grass-like
P pinnate	L linear
B bipinnate	N lanceolate
M palmate	E elliptic
T trifoliate	O ovate
	C hastate, cordate
	R 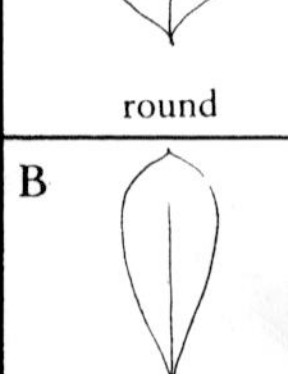 round
	B obovate

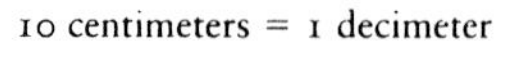
10 centimeters = 1 decimeter

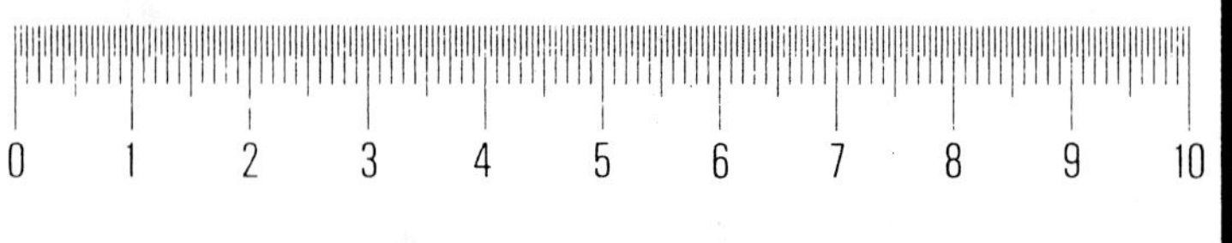

	Leaf Margin
E 	entire
S 	serrate
D 	dentate
C 	cut
L 	lobed

Key Flower Characters					
Flower Arrangement		Flower Form		Flower Color	
S 	solitary	N	none, or not evident	G	green
I 	spike	R 	round flat	W	white, cream
R 	raceme	C 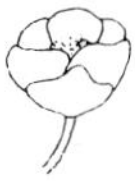	cup	Y	yellow, orange
P 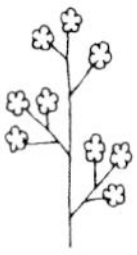	panicle	F 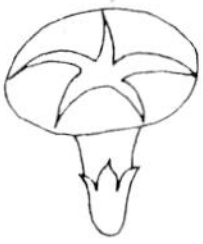	round funnel	R	pink, red
U 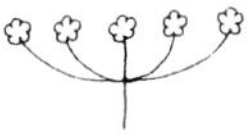	umbel, corymb, cyme	T 	round tube	B	blue, purple
H 	head & rays	Z 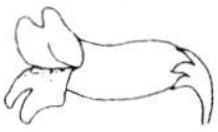	two lipped, zygomorphic	M	maroon, brown
K 	knot, glomerule, head				
T 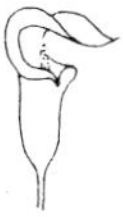	spathe				

Wild Flowers *of* North Carolina

Wild Flowers *of* North Carolina

SECOND EDITION

WILLIAM S. JUSTICE,

C. RITCHIE BELL, &

ANNE H. LINDSEY

The University of North Carolina Press CHAPEL HILL & LONDON

The publisher and authors gratefully acknowledge the contributions to this work of the North Carolina Botanical Garden and its programs for the conservation and propagation of native plants in North Carolina.

Set in Scala and The Sans types
by Eric M. Brooks
Manufactured in China

The paper in this book meets the guidelines for permanence and durability of the Committee on Production Guidelines for Book Longevity of the Council on Library Resources.

Library of Congress
Cataloging-in-Publication Data
Justice, William S.
Wild flowers of North Carolina /
William S. Justice, C. Ritchie Bell, and
Anne H. Lindsey—2nd ed.
p. cm.
Includes bibliographical references and index.
ISBN 0-8078-2933-1 (cloth: alk. paper)
ISBN 0-8078-5597-9 (pbk.: alk. paper)
1. Wild flowers—North Carolina—Identification. 2. Wild flowers—North Carolina—Pictorial works. I. Bell, C. Ritchie II. Lindsey, Anne H. III. Title.
QK178.J8 2005
582.13'09756—dc22 2004019095

cloth 09 08 07 06 05 5 4 3 2 1
paper 09 08 07 06 05 5 4 3 2 1

Frontispiece: Venus Flytrap
(*Dionaea muscipula*)

To the memory of

WILLIAM S. JUSTICE, M.D.,

surgeon, photographer,

botanist, friend

CONTENTS

ACKNOWLEDGMENTS

A generation has passed since a group of wonderful, talented people across North Carolina provided the strong "support group" that made possible the production and publication of the first edition of *Wild Flowers of North Carolina*, which has sold over 100,000 copies and obviously provided the foundation upon which this enlarged second edition has been built. Thus we say thanks again to these "old timers," many of whom have passed on while their legacy continues to inspire: Dr. John Barber, Gladstone McDowell, Mrs. John DuBose, Mr. and Mrs. Thomas Shinn, Mrs. Ruth Landolina, Dr. Robert L. Wilbur, Dr. Albert E. Radford, and a generation of members of the Garden Club of North Carolina.

The authors wish to acknowledge with many thanks Dr. Peter White, director of the North Carolina Botanical Garden; Alan Weakley, curator of the University of North Carolina Herbarium; Carol Ann McCormick, assistant curator of the University of North Carolina Herbarium; "Rusty" Russell of the Smithsonian Institution, for assistance in providing copies of some of William Justice's slides; and, of course, the staff of the North Carolina Botanical Garden, who were essentially "on call" for help with any botanical or conservation questions during our work on this second edition.

INTRODUCTION

Many of the world's wild flowers are fast disappearing as habitat destruction accelerates with exponential human population growth. There is, however, cause for hope that we may be able to halt or reverse this trend. Awareness is the first step. Our goal in this presentation of 500 native or naturalized plants in North Carolina is to open eyes, minds, and hearts to the story of the state's wild flowers, their beauty, their interesting attributes, their uses, and, in many cases, their plight. We hope that by stimulating a greater interest in this beautiful and unique natural resource we will also increase awareness of the need for, and value of, its preservation. Anyone who takes the time to get to know and appreciate even one of our wild flowers helps to preserve them.

Since the publication of the first edition of *Wild Flowers of North Carolina* in 1968, four major developments have changed the landscape for plant conservation. The first of these is a growing awareness of the importance of species diversity, habitat conservation, and the need to mitigate for the harm done by development, as exemplified in passage of the Endangered Species Act. Second is an increased interest in growing native plants that are well adapted to the climates and soils in which they grow and better able to survive local variations in rainfall or temperature. Third, as a direct consequence of the first two developments, recent years have seen the advent of true native plant nurseries—nurseries that supply gardeners' and landscapers' demand for native plants by propagating them from seeds, cuttings, or tissue culture in the nursery rather than collecting plants in the wild. Finally, there has been a renewed interest in the indigenous use of native plants for medicine. Exploration of this "green pharmacy" has led to the development of modern medicines as well as an increase in the use of traditional herbal remedies.

The Endangered Species Act of 1973 began a concerted effort to protect some of our rarest plants. In 1974 The Nature Conservancy helped to establish the first state natural heritage program, a program that now exists in all 50 states. The North Carolina Natural Heritage Program is administered by the Office of Conservation and Community Affairs within the North Carolina Department of Environment and Natural Resources. The program inventories, catalogues, and facilitates protection of the rarest and the most outstanding elements of the natural diversity of our state. Legal protection and categorization of rare plants falls to the North Carolina Department of Agriculture Plant Conservation Program formed under the Plant Protection and Conservation Act of 1979. This program has responsibility for the legal protection of plants in three categories (endangered, threatened, or of special concern), enforces regulations and issues permits concerning state-listed plant species, monitors and manages populations of listed species, provides educational materials to the public, and monitors trade in American Ginseng. A list of state and national conservation organizations is provided in Table 1.

If given a place to grow, with adequate light, water, and soil conditions, many of the 3,500 or more native plants that once formed what B. W. Wells called the "natural gardens of North Carolina" can be as colorful and interesting as the "usual horticultural suspects," if not more so. The North Carolina Botanical Garden initiated the very important concept of "conservation through propagation" in 1978, promoting "appropriate" propagation of native plants for use in the horticultural trade. Its "plant of the year" program has introduced 20 native plant species into the trade, and native plant nurseries that propagate rather than collect the plants they sell are now abundant in North Carolina and the rest of the country.

Alternative medicine is increasingly being incorporated into modern Western medicine, leading to a burgeoning interest in native plants for their medicinal potential. The use of plants in medicine was commonplace until the mid-nineteenth century, when the promise of chemistry and technology led to the rejection of homeopathic remedies for a variety of illnesses. Now the plants that were used by Native Americans and by early colonists, who often brought plants with them for their medicinal needs, are again the object of study. A growing number of plants have recently been found to have modern medical

TABLE 1. SOME STATE AND NATIONAL ORGANIZATIONS DEDICATED TO THE CONSERVATION OF NATIVE PLANTS

Organization	*Purpose*	*Web Site*
STATE		
The Plant Conservation Program	Promotes the legal protection of plants that are endangered, threatened, or of special concern	www.ncagr.com/plantind/plant/conserv/cons.htm
North Carolina Heritage Program	Inventories, catalogues, and facilitates protection of the state's rarest plants	www.ils.unc.edu/parkproject/nhp/
North Carolina Botanical Garden	Promotes knowledge, appreciation, and conservation of North Carolina's flora through public displays, education outreach, research, and public programs linked to the garden's "conservation through propagation" theme	www.ncbg.unc.edu/
North Carolina Wild Flower Preservation Society	Promotes the enjoyment and conservation of native plants and their habitats through education, protection, and propagation	www.ncwildflower.org/
NATIONAL		
Plant Conservation Alliance (PCA)	A consortium of 10 federal government agencies and over 145 nonfederal groups (representing various disciplines within the conservation field) that work collectively to solve the problems of native plant extinction and native habitat restoration, ensuring the preservation of our ecosystem	www.nps.gov/plants
Center Plant for Conservation (CPC)	A network of more than 30 leading botanic institutions, founded in 1984, dedicated solely to preventing the extinction of very rare U.S. native plants	www.centerforplantconservation.org

uses, among them Mayapple (for combating testicular cancer), Wild Yam (for the production of steroid drugs), and Bloodroot (as a source for the plaque-fighting sanguinarine). There is great promise for more such discoveries. Herbal remedies newly catalogued can be found in Dr. James Duke's *Green Pharmacy* (see References).

Another important change that has taken place over the past 30 years is the revision of the scientific names given to some of the species of wild flowers found in North Carolina. In some cases, plants previously considered as two separate species have been combined into one; in other cases, a species that was once viewed as a single entity has been divided into two species. Most often, close studies of plant groups have revealed that they need to be placed in different genera with new names. Appendix 3 lists all nomenclatural changes relevant to the wild flowers of North Carolina that have occurred since this book's original publication.

Format

In the limited space available for the text material associated with each of the 500 plants illustrated, an attempt has been made to cover the following specific items of pertinent information in a relatively uniform sequence and format most useful to the person interested in the flowering plants of North Carolina and the surrounding regions:

1. common name
2. scientific name
3. whether the plant is native or introduced
4. size perspective
5. general comments on interesting aspects of the plant
6. frequency, especially if very rare or very common
7. habitat
8. range in North Carolina
9. general range in the United States
10. months in bloom in our area
11. index reference number from the *Manual of the Vascular Flora of the Carolinas*
12. key character summary code
13. cultivation, medicinal, poisonous, and plant status codes

Since each of the brief entries is independent of the others, and since each entry can be read in only a few seconds, the information concerning items 3 through 9 will not always be in the exact sequence given above, varying as seems appropriate for easy reading—and for what is visually evident (e.g., flower color) in the photograph. Some comments concerning the application of each piece of information are given, by category, in the following paragraphs.

COMMON NAME

If a native plant of North America had any resemblance to one in Europe, the early colonists often applied the European common name to the New World plant even though the two may have been completely unrelated botanically. Depending upon the country or area of origin of the colonists, one particular plant might have several common names in different parts of this country. Or a given plant might have two or more common names in a single area because of different aspects of its appearance or use: thus the attraction of the colorful flowers of *Asclepias tuberosa* for butterflies accounts for the common name "Butterfly Weed," while an old medicinal use accounts for the common name "Pleurisy-Root." On the other hand, a single name might be applied to a number of different plants: the common name "Buttercup," for example, has been given to plants belonging to several of the yellow-flowered species of *Ranunculus* (in the family Ranunculaceae) and also to several species of the completely unrelated genus *Narcissus* (in the family Amaryllidaceae). Common names are sometimes easier to remember than the scientific name but they are certainly not exact! Furthermore, many common names are the same as the generic name or the first part of the scientific name, such as Rhododendron, Iris, Trillium, Magnolia, Sassafras, and Oxalis, and no one ever thinks of these names as being too hard to learn.

SCIENTIFIC NAME

The scientific name of a plant consists of two Latin or latinized words, a genus or *generic* name followed by a species name or *specific epithet.* By international agreement on the rules governing the formation of scientific names of plants, no two kinds may have the same name; thus every kind of plant has a different combination of generic name and specific epithet. Although a specific epithet may be repeated from one genus to the next (e.g., *Magnolia virginiana* for Sweet Bay and *Fragaria virginiana* for the Wild Strawberry), the generic names are different and indicate that the plants are different—in this case they even belong to different plant families, the Magnolia family (Magnoliaceae) and the Rose family (Rosaceae) respectively.

In cases where one plant has been given different scientific names by different botanists, the variance can be explained in one of two ways. Since 1930 considerable agreement has been reached, on an international level, concerning the rules for naming plants. Although aimed at ultimate stability, the new international rules have necessitated many changes in names that were used by botanists in this country for the 200 years before 1930—when many of our native plants were first discovered and named. Those earlier names that have been replaced are now officially invalid. Another situation, involving botanical opinion rather than rules, often occurs if a species is quite variable or is poorly known. Under such conditions, different acceptable botanical interpretations of the presumed relationships result in two concepts and thus two names. Further taxonomic research might shed more light on the patterns of plant variation and relationship and ultimately resolve the problem. Even so, botanical names are far more uniform and stable, the world over, than common names, and therefore they are given here, as are the scientific family names, to aid those who might wish to search further for information on a particular plant species of special interest.

The scientific name is followed by the name of the botanist who first described and classified the plant. This latter item may be either a name or an abbreviation and may consist of one or two names, the first of which is then in parentheses. This indicates that a second botanist brought about some change in the status of the botanical name after it was first applied by the original author, whose name then appears in the parentheses. These author, or authority, names are important to botanists as bibliographic references. For example, "Walter" after a plant name indicates that the plant was named by Thomas Walter, whose classic *Flora Caroliniana* was published in 1788. In a similar way, "Linnaeus" or "L." after a plant name indicates that the plant was named by the Swedish botanist Carolus Linnaeus, who first published many plant names and descriptions in his *Species Plantarum* in 1753. The large number of our plants first described by Linnaeus (1707–78), William Bartram (1739–1823) of Pennsylvania, Thomas Walter (1740–89) of South Carolina, and the French botanist-explorer André Michaux (1746–1802) reflects the extensive early botanical exploration of eastern North America in general and of the Carolinas in particular.

NATIVE VS. INTRODUCED

In most cases the botanical literature is complete enough to show which plants now growing in our area without cultivation are truly native and which have been introduced from other areas, chiefly Europe. This information is given, in either direct or indirect form, for each species. It is often coupled with information on the life span of the plant: whether it is *annual* and lives only one year or one season, *biennial* and lives for two years (usually blooming the second year, when it then sets seeds and dies), or *perennial* and lives for three or more years. Recent emphasis has been paid to the widespread destructive effect of exotic invasives, those introduced nonnative species, often perennials, that aggressively overtake natural habitats and displace native plants. Invasive plants are identified in the text.

SCALE OR PLANT SIZE

In each entry, reference is made to the size of the entire plant, or to some specific part, in order to give an idea of the scale of the picture and to aid in identification of the plant. These measurements are helpful, and sometimes critical, in identification because similar and closely related species may differ only in some aspect of size. As is now usual in botanical references, and indeed in scientific fields worldwide, all measurements are given in metric units. This decimal-based international system measures from 1/1000 of a meter (a millimeter) to 1/100 of a meter (a centimeter) to 1/10 of a meter (a decimeter) to a meter itself and allows for much greater flexibility and accuracy than is possible in the English system. However, to help picture the actual size and to assist with the conversion from English to metric, a decimeter rule, marked in millimeters, is provided at the front and back of the book. To make the switch to metric, there are four ready references to keep in mind: a millimeter (mm) is approximately the thickness of a penny; 2.5 centimeters (cm), or 25 mm, are approximately 1 inch; a decimeter (dm), or 10 cm, is approximately 4 inches; a meter (m), or 10 dm, is just a bit more than 3 feet. The sources for size information for each species are the *Manual of the Vascular Flora of the Carolinas* and the *Manual of Vascular Plants of Northeastern United States and Adjacent Canada* (see References).

GENERAL COMMENTS

We include brief comments on matters of particular interest, as appropriate for the plants illustrated. These may touch on the medicinal or other uses made of the species, horticultural information, and interesting biological facts such as what pollinates the flowers or distributes the seed. Comments are also made regarding other closely related, and often similar, species found in North Carolina.

FREQUENCY

Information on frequency (common, frequent, rare), though subjective and difficult to apply in many particular cases, is nonetheless included as a relative guide. This knowledge is of particular importance in respect to our rare native plants that are in danger of becoming extinct. As relevant, plant status (whether it is endangered, etc.) is also indicated by a code in the last line of the entry (see below).

HABITAT

Usually each of our native wild flowers grows only in the particular habitat or range of habitats to which it is adapted. In order to help you know where to look for certain plants and to aid in the identification of others that you may find unexpectedly, the habitat is given for each species treated, as listed in the *Manual of the Vascular Flora of the Carolinas.*

RANGE IN NORTH CAROLINA

Varying environmental conditions in North Carolina's three geographic provinces—mountains, piedmont, and coastal plain (see Figure 1)—also influence plant distribution. As shown in Table 2, some species are limited to one or two of these provinces, while others may be found throughout the state. For this reason, knowledge of a plant's usual range may aid in the discovery and identification of some of the wild flowers treated here. If an entry reads "chiefly mountains," it means that most of the known localities for this species in our state are in the mountains but that a few are also known from the piedmont; "chiefly coastal plain and piedmont" would indicate primary distribution in these provinces but with a few localities in the mountains; "scattered throughout" means just that—specimens are found here and there, in the appropriate habitat, in each of the three provinces.

GENERAL RANGE IN NORTH AMERICA

To illustrate the diversity of the regions of origin of our plants and to make this book more useful to those in areas adjacent to North Carolina, some indication is given of the general range of each species considered. These ranges are given as specifically as possible but are sometimes generalized for economy of words. For example, "ranging throughout the United States" does not necessarily mean that the species occurs in *every* state, "eastern and central United States" generally means the eastern half of the United States, and "eastern North America" means the species extends at least into southern Canada.

MONTHS IN BLOOM

The time at which each plant is normally in bloom in North Carolina is listed as a range of months. "May–June" would indicate that the plant blooms in our area during part or all of May and June. Of course, for plants of a species found throughout the state, those in the mountains will usually bloom later than those at lower elevations.

INDEX NUMBER

For those who might desire more specialized or detailed information on a plant or its relatives, the set of three numbers in parentheses fol-

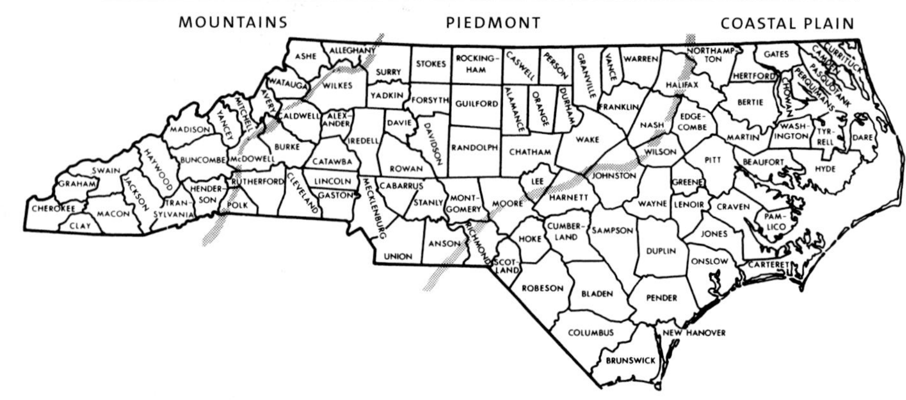

FIGURE 1. *Counties and provinces of North Carolina.*

lowing the months in bloom is provided as an index to each species in the *Manual of the Vascular Flora of the Carolinas* (see References). For example, the index number of Black Locust, 98-32-1, refers to the *Manual*'s family number 98 (Fabaceae), genus 32 (*Robinia*) within this family, and species number 1 (*Robinia pseudoacacia*) of this genus. The sequence of the plants illustrated in this book thus follows the botanical sequence, based on general relationships, that is found in the *Manual*.

The information on distribution in North Carolina, blooming dates, and habitats is taken primarily from the *Manual of the Vascular Flora of the Carolinas*, supplemented by nearly a century of combined field observations on the part of the authors. General ranges are derived from both the U.S. Department of Agriculture "Plants Database" (plants.usda.gov) and the *Manual of Vascular Plants of Northeastern United States and Adjacent Canada* (see References).

Because there have been changes in the taxonomy of several species since the *Manual of the Vascular Flora of the Carolinas* was published, we have used the new scientific name in the treatments of those species but maintain the old index number. A list of all name changes appears in Appendix 3. Additionally, for those plants that have undergone a name change, the now-superseded name appears in brackets after the current name at the beginning of the entry.

TABLE 2. DISTRIBUTION BY PROVINCE AND RELATIVE FREQUENCY OF THE 2,945 SPECIES OF FLOWERING PLANTS OF NORTH CAROLINA[a]

Number of Species Found In:	*Three or Fewer Counties*	*Four or More Counties*	*Total Species*
Mountains Only	171	142	313
Both Mountains and Piedmont	31	313	344
Piedmont Only	138	45	183
Both Piedmont and Coastal Plain	55	516	571
Coastal Plain Only	206	263	469
Both Coastal Plain and Mountains	8	26	34
Throughout	6	1,025	1,031
Totals	615	2,330	2,945

a. Based on data from Radford, Ahles, and Bell, *Manual of the Vascular Flora of the Carolinas* (1968). Exact numbers today will differ, but proportions will be similar.

KEY CHARACTER SUMMARY CODE

On the next-to-the-last line of each entry, we have included a morphological code that summarizes several characters of the plant according to the pictorial key located at the front and back of the book. The first three code elements are important general plant characters: habit, petal number (includes petal-like sepals), and leaf arrangement. The second grouping concerns specific important leaf characters: type, shape, and margins. The final three code elements refer to information on flower characters: arrangement, form, and color. The values for each of these characters has necessarily been simplified for ease in categorization, but the code can be useful when trying to determine the identity of a new or unknown wild flower. For detailed examples of how the code works, see Table 3.

CULTIVATION, MEDICINAL, POISONOUS, AND PLANT STATUS CODES

The last line of each entry provides symbols and codes relaying information about a plant's soil and water requirements and indicating, as relevant, if it is available in the trade, if it has medicinal uses, if it's poisonous, and if (as well as to what degree) it is rare. An explanation of the symbols and codes follows.

TABLE 3. KEY CHARACTER SUMMARY CODE EXAMPLES

EXAMPLE 1		EXAMPLE 2	
Rue Anemone		Yellow Jessamine	
Thalictrum thalictroides		*Gelsemium sempervirens*	
H-VA/BOE/URW		V-50/SNE/SFY	
H	terrestrial herb	V	vine
V	variable petal number	5	5 petals
A	alternate leaves	O	opposite leaves
B	bipinnate leaf type	S	simple leaf type
O	ovate leaflet shape	N	lanceolate leaf shape
E	entire leaflet margin	E	entire leaf margin
U	umbellate flower arrangement	S	solitary flower arrangement
R	flowers round, flat	F	flowers funnel shaped
W	flowers white	Y	flowers yellow

Light Requirements

❍ full sun

● full shade

◗ partial shade and sun or filtered sun all day

Soil Moisture

A aquatic

W habitat wet most of the year

M moist, soil should not dry out

N average, or variable soil moisture

D dry soils

Cultivation

$ This symbol is used when there is clear evidence of general availability of the species in the trade—either seeds or nursery-grown plants, or both. (Note: When you buy plants, be sure they are indeed nursery-propagated and -grown and not wild "gathered." The former usually live and thrive; the latter usually don't!)

Medicinal Use

○ This symbol is shown when there is a record of medicinal use for the plant in either *A Field Guide to Medicinal Plants and Herbs of Eastern and Central North America* by Steven Foster and James A. Duke or *Native American Ethnobotany* by Daniel E. Moerman (see References).

Poisonous Plants

◉ poisonous when ingested

△ poisonous when ingested but either of low toxicity or requires eating large quantities to achieve toxicity

☞ toxicity is external, causing a skin reaction

Information on plant toxicity is derived from the online database *Poisonous Plants of North Carolina* (see References).

Plant Status

X This symbol, indicating "Do Not Dig!," appears at the end of the general comments and range indications, before the bloom months, for plants that should never be transplanted from the wild because they won't survive or because they are very rare.

E endangered: species in jeopardy

T threatened: species likely to become endangered in the foreseeable future

SC endangered and threatened: species listed as "of special concern," which may be propagated and sold under permit from the North Carolina Department of Agriculture

SR very rare in North Carolina—may be more common elsewhere in its range

R rare in North Carolina based on the authors' field experience—either located in only a few locations or generally uncommon throughout its range in the state, but without state or federal status

P parasitic

M mutualistic relationship with fungal partners (mycorrhizae)

S saprophytic

APPENDIXES

Following all of the entries for specific plants are three appendixes. The first is a horticultural chart for all native wild flowers of North Carolina. It presents much of the same information that can be found in specific entries (bloom time, soil requirements, etc.) but in an abbreviated, easy-to-consult form and with the plants listed in alphabetical order (by scientific name). Appendix 1 also indicates the references that are the source of information on medicinal uses and propagation found in individual plant entries.

Appendix 2 provides a list of North Carolina wild flowers that are classified as endangered or threatened by state and/or federal agencies. Appendix 3 lists nomenclatural changes affecting the state's wild flowers since publication of the first edition of this book.

References

Bell, C. Ritchie. 1967. *Plant Variation and Classification.* 135 pp. Belmont, Calif.: Wadsworth.

Dirr, Michael A., and Charles W. Heuser Jr. 1987. *The Reference Manual of Woody Plant Propagation: From Seed to Tissue Culture.* 239 pp. Athens, Ga.: Varsity Press.

Duke, James A. 1997. *The Green Pharmacy.* 508 pp. Emmaus, Pa.: Rodale Press.

Flora North America Online: Flora of North America Editorial Committee, eds. 1993– . *Flora of North America North of Mexico.* 7 vols. to date. New York: Oxford University Press.

Foster, Steven, and James A. Duke. 2000. *A Field Guide to Medicinal Plants and Herbs of Eastern and Central North America.* 2nd edition. 411 pp. Boston, Mass.: Houghton Mifflin.

Gleason, Henry A., and Arthur Cronquist. 1963. *Manual of Vascular Plants of Northeastern United States and Adjacent Canada.* 810 pp. Princeton, N.J.: Van Nostrand.

Moerman, Daniel E. 1988. *Native American Ethnobotany.* 927 pp. Portland, Ore.: Timber Press.

North Carolina Natural Heritage Database (www.ncsparks.net/nhp/search.html). The North Carolina Natural Heritage Program is a part of the Office of Conservation and Community Affairs within the North Carolina Department of Environment and Natural Resources.

Phillips, Harry R. 1985. *Growing and Propagating Wild Flowers.* 331 pp. Chapel Hill: University of North Carolina Press.

Radford, Albert E., Harry E. Ahles, and C. Ritchie Bell. 1968. *Manual of the Vascular Flora of the Carolinas.* lxi, 1183 pp. Chapel Hill: University of North Carolina Press.

Russell, Alice B., James W. Hardin, Larry Grand, and Angela Fraser. *Poisonous Plants of North Carolina.* Online database (www.ces.ncsu.edu/depts/hort/consumer/poison/poison.htm). Departments of Horticultural Science, Botany, Plant Pathology, and Family and Consumer Sciences, North Carolina State University, Raleigh. Programming, Miguel A. Buendia; Graphics, Brad Capel.

U.S. Department of Agriculture, Natural Resources Conservation Service. *The Plants Database,* Version 3.5 (http://plants.usda.gov). National Plant Data Center, Baton Rouge, La.

Flower Structure and Function

Not all plants have flowers. The lower or more primitive types—algae, fungi, mosses, and ferns—do not produce flowers or seeds in their reproductive process but produce microscopic spores that produce (or become!) gametes. Pines and their relatives are wind pollinated and do form seeds but still have no structure that can be correctly called a flower. Only the more highly evolved plants—known botanically as the Angiosperms—have the specialized, complex, and often beautiful structures we call flowers associated with the sexual reproductive process that produces seeds.

A typical flower (Figure 2) is made up of four sets of parts, arranged in whorls or in concentric rings. Each part of the flower is specialized for a particular function. The outermost series of parts is the *calyx,* which is made up of the *sepals.* The calyx, in which the individual sepals may be entirely separate or fused to varying degrees, protects the flower in bud, and is usually green but, as in many lilies, may be as brightly colored as a petal and thus function in the attraction of the insect (or other animal) pollinators. The second whorl or series of parts is the *corolla,* which is made up of the *petals,* which also may be either

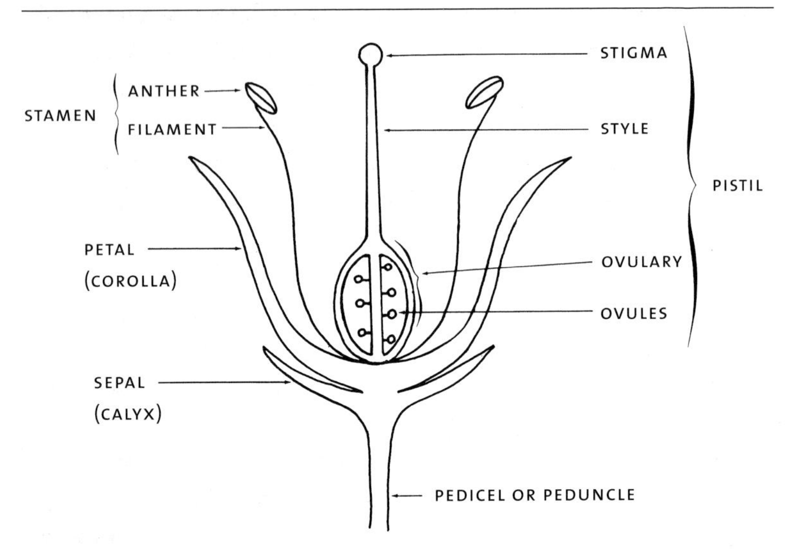

FIGURE 2. *The parts of a typical flower.*

fused, as in the Morning Glory and Rhododendron, or separate, as in the Lily and Cactus. The corolla is usually the most colorful part of the flower, the color, shape, and scent serving to attract the insects or birds often necessary for *pollination,* which is the transfer of pollen from the anthers to the stigma. The third whorl is composed of the *stamens,* each of which bears a pollen sac, or *anther,* at the end of a slender *filament.* The pollen produces the cells that function as the male gametes. At the center of a typical flower is the *pistil,* or female reproductive portion, made up of three parts: the pollen-receptive area, the *stigma;* a region of connective tissue, the *style;* and the ovule producing portion, the *ovary* or *ovulary.* The ovules, after fertilization, will develop into *seeds,* and the ovulary will become the *fruit.*

Not all flowers, however, have all of these parts. Some, for example, have no petals. In these cases the flowers are frequently small, insignificant, and not very pretty. Such flowers would offer little attraction to pollinating insects. Thus it is not surprising to find that the pollen of these *apetalous* flowers is spread by the wind. Grasses, many trees (such as the oaks and willows), and even some plants in the Aster family (such as the familiar Ragweed) lack showy petals and are wind pollinated. On the other hand, some plants having flowers without petals or with insignificant ones are nonetheless insect pollinated. When this is true, other structures have usually been modified through change in color and shape to function as "petals." Such is the case with Clematis, in which the sepals are colorful and petal-like, and with Dogwood, in which we find that the "flower" is actually made up of a cluster of small inconspicuous flowers surrounded by four showy white modified leaves or *bracts*. Another type of "flower" is found in such plants as Skunk Cabbage and Jack-in-the-Pulpit. In the case of these two, many small flowers are borne on a fleshy stalk, called the *spadix*, surrounded by a colorful envelope of tissue called the *spathe*. The color, shape, and odor of the spathe attract the pollinators.

A flower with both stamens and pistils, as in Figure 2, is said to be *perfect*. However, the flowers of some plants are found to be unisexual; that is, they may be either *staminate*, with stamens only, or *pistillate*, with pistils only. If both male and female, or staminate and pistillate, flowers are found on a single plant, it is said to be *monoecious*. If the male flowers are produced on one plant and the female flowers on another, as in the case of Holly, the plants are said to be *dioecious*. Most of the wind-pollinated plants mentioned earlier can be either monoecious or dioecious and produce many more male than female flowers. The long clusters of male flowers, called catkins, are conspicuous on our alders, oaks, and hickories in early spring; the relatively few, small female flowers in the leaf or bud axils are rarely noticed.

Flowers are modified through adaptive changes in the shape, size, and relationship of parts as well as through the loss of parts. Flowers in the Lily family, Liliaceae (Figure 3a); the Rose family, Rosaceae (Figure 3b); and the Potato family, Solanaceae (Figure 3c) are *actinomorphic*. That is, they are quite regular or radially symmetrical. In contrast, the flowers in the Orchid family, Orchidaceae (Figure 4a); the Legume family, Fabaceae (Figure 4b); and the Mint family, Lamiaceae (Figure 4c) typically offer examples of *zygomorphic* flowers, or flowers that have a bilateral (often two lipped) symmetry.

Inflorescence is used as a general term to indicate the specific flower-bearing portion of the plant and usually refers to the particular type of arrangement of the flowers on the stems and branches (Figure 5).

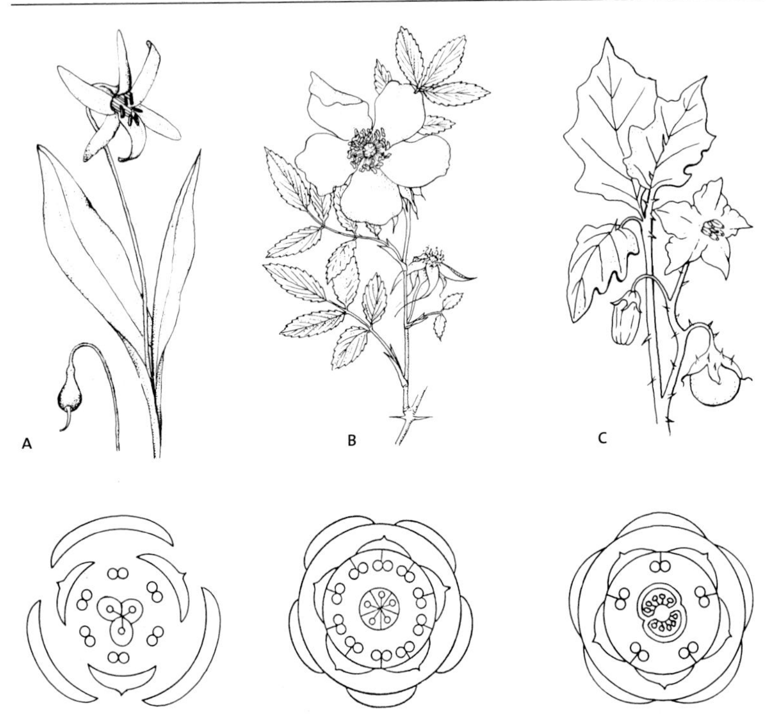

FIGURE 3. *Examples of flowers with radial symmetry. The "floral diagrams" below each drawing represent schematically a cross section of the flower shown as representative for the family. The diagrams show the characteristic number and arrangement of the floral parts (sepals, petals, stamens, pistil) in each whorl of the flower. Such diagrams are often useful aids in plant identification. (a) Trout Lily (*Erythronium americanum*) Lily family (Liliaceae); a monocotyledon with parallel venation in the leaves and flower parts in sets of 3; the leafless flowering stalk of this plant is a good example of a scape. (b) Carolina Rose (*Rosa carolina*) Rose family (Rosaceae); a dicotyledon with net-veined leaves and flower parts in sets of 5, except for the fact that in this family there are usually many stamens; note the pinnately compound leaves with stipules at the base. (c) Horse Nettle (*Solanum carolinense*) Potato family (Solanaceae); a dicotyledon; the floral diagram indicates the sepals are fused, the petals are fused, and the stamens arise from the fused lower part of the corolla. (From Bell,* Plant Variation and Classification, *used by permission)*

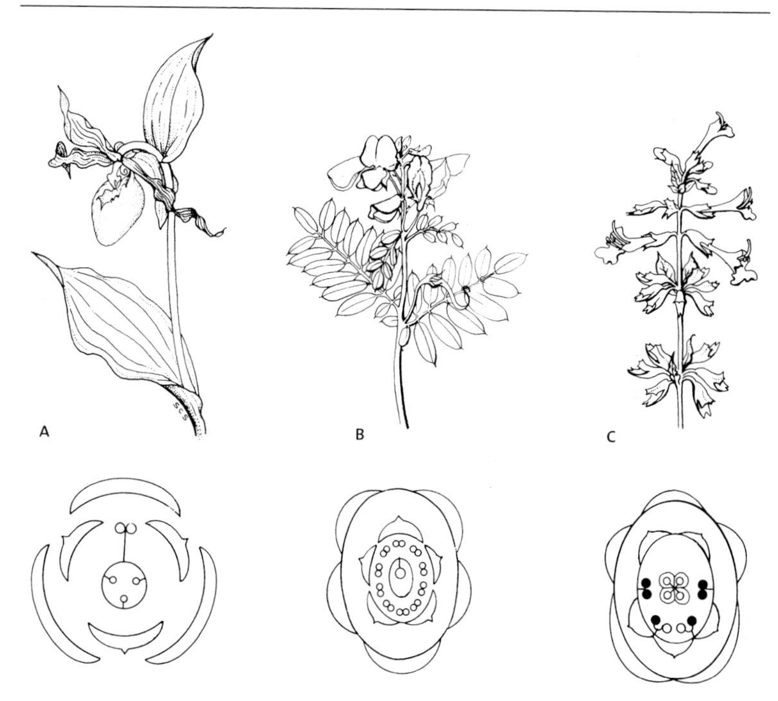

FIGURE 4. *Examples of flowers with bilateral symmetry. The "floral diagrams" reflect the zygomorphic nature of the flowers, as well as the number and arrangement of the flower parts. (a) Yellow Lady Slipper (*Cypripedium calceolus*) Orchid family (Orchidaceae); another monocotyledon with typical parallel veins in the leaves and flower parts in sets of 3, but note in the diagram the single stamen characteristic of the family fused to the pistil. (b) Goat's Rue (*Tephrosia virginiana*) Bean family (Fabaceae); a dicotyledon with reticulate veins in the leaflets of the pinnately compound leaves and with flower parts in sets of 5. (c) Lyre-leaved Sage (*Salvia lyrata*) Mint family (Lamiaceae); a dicotyledon; the sterile stamens of this plant are shown in black in the diagram. (From Bell,* Plant Variation and Classification, *used by permission)*

The "flower" of the Jack-in-the-Pulpit, for example, is really a group of flowers in a specialized arrangement or inflorescence, and in the Aster family we also find that many "flowers," such as the common Daisy and the Black-eyed Susan, are not single flowers at all but are groups of flowers in a compact inflorescence or head. Furthermore, such heads often contain both actinomorphic and zygomorphic flowers. Each "petal" of a daisy is actually a separate flower (called a *ray* flower),

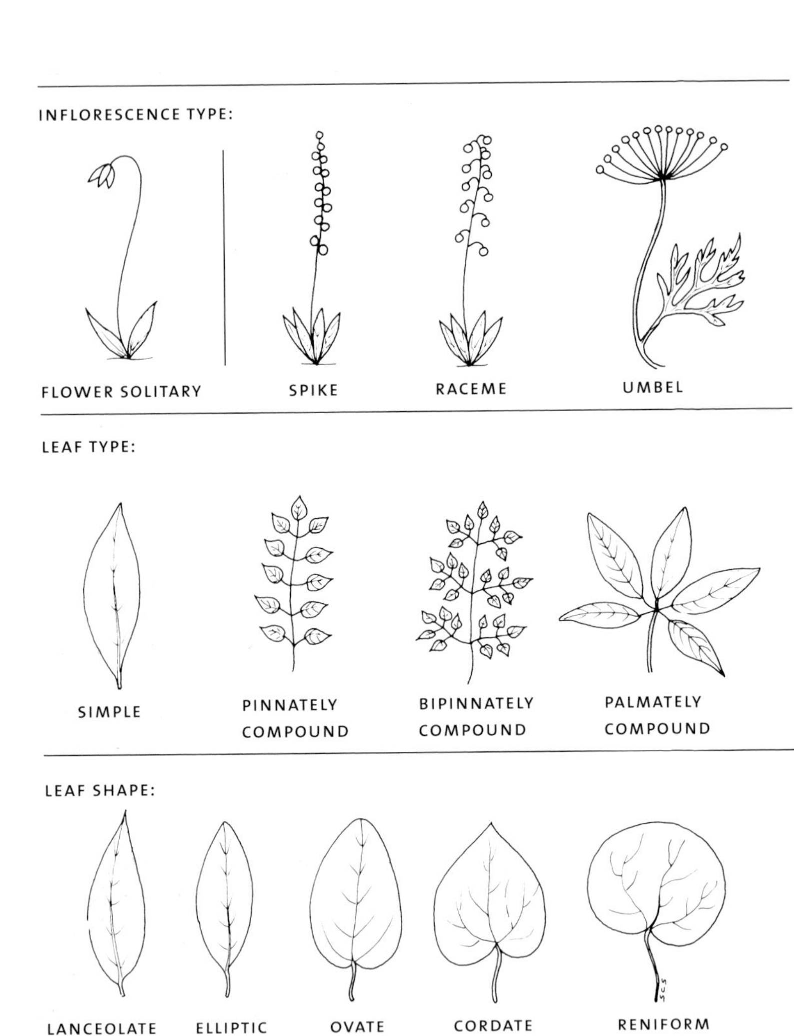

FIGURE 5. *Inflorescence and leaves.*

with one large white petal. The center of the daisy is made up of many regular, small, yellow flowers (called *disc* flowers) with uniform petals (Fig. 6).

Flowering plants are divided into two major groups on the basis of minute internal differences, but members of each group usually can be recognized on the basis of flower structure, or flower structure in association with leaf structure. Monocotyledons (plants with only a single

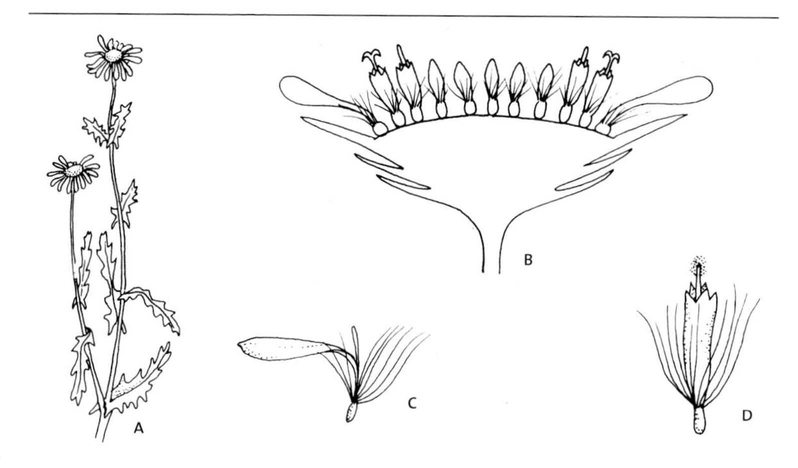

FIGURE 6. *Flowers of the Aster family (Asteraceae). (a) Habit sketch of Ox-eye Daisy (*Leucanthemum vulgare*). (b) Diagrammatic section through a daisy head or inflorescence showing the two different kinds of flowers: ray flowers (c) are the zygomorphic outer flowers or "petals" of the daisy, and disc flowers (d) are the radially symmetrical inner flowers that form the yellow center of the inflorescence. In (b) the disc flowers are shown in different stages of development; note the bracts shown on the receptacle just below the flowers. (From Bell,* Plant Variation and Classification, *used by permission)*

embryonic leaf, such as the lilies, orchids, and grasses) generally have flower parts in threes and leaf veins that are more or less parallel; dicotyledons (plants with two embryonic or seed leaves, such as the mustards, asters, and mints) generally have flower parts in fours or fives or in multiples of four or five and leaf veins that are reticulate, or netlike, and not parallel. In a similar way, different combinations of structural patterns enable us, with practice and careful application, to distinguish individual plants and to assign them to their proper families. The variations in structure between unrelated plants and the similarities within groups are used by botanists in both identification and classification.

Plants of a given area can be identified by the use of a "key" or orderly list of contrasting characters by means of which the plant under consideration may be assigned to smaller and smaller groups until it is ultimately identified. However, a "key" for only 500 of our nearly 3,000 flowering plants would be inaccurate and misleading. For that reason, in addition to the deliberate avoidance of a technical approach in this presentation, such a device has been omitted. Complete keys for the identification of all of our flowering plants can be found in the *Manual of the Vascular Flora of the Carolinas*, and, as noted above, each species illustrated in this book is coordinated by reference number to the more detailed material in the *Manual*.

Glossary

Actinomorphic: Regular, radially symmetrical; descriptive of a flower, or of a set of flower parts, that can be cut through the center into equal and similar parts along two or more planes.

Acuminate: Long-tapering to a pointed apex.

Alternate leaf arrangement: A single leaf per node.

Angiosperm: One of the flowering plants (ovules enclosed in an ovulary).

Annual: Life cycle completed in one year or less.

Anther: The fertile part of the stamen; the part that produces the pollen.

Auriculate: An ear-shaped appendage or lobe.

Axil, leaf: The angle formed between the stem and the upper surface of the leaf.

Axillary: In an axil.

Basal: At the base of the stem.

Berry: A simple, fleshy fruit in which the ovulary wall remains succulent.

Biennial: A plant that completes its life cycle in two years and then dies; normally remains as a vegetative rosette the first year, flowering the second year.

Bilabiate: Two-lipped, as some corollas that are thus zygomorphic.

Blade: The expanded or flattened part of the leaf.

Bract: A reduced leaf, particularly one subtending a flower, or inflorescence, as the involucral bracts in the Aster family.

Bud: An aggregation of undeveloped leaves, or flowers, or both, on an axis with undeveloped internodes, often enclosed by scales.

Bulb: A short underground stem surrounded by fleshy leaves or scales.

Campanulate: Bell-shaped, as some corollas.

Capsule: A dry, dehiscent fruit derived from two or more united carpels.

Catkin: A scaly bracted, usually flexuous spike or spikelike inflorescence, often of unisexual flowers.

Cauline: Occurring along a stem—as opposed to basal.

Cespitose: Tufted in clumps.

Chasmogamous flower: A normal open flower, not cleistogamous.

Cleft: Cut 1/4 to 1/2 the distance from the margin to midrib, or apex to base, or, generally, any deep lobe or cut.

Cleistogamous flower: A self-pollinated flower that does not open.

Colonial: A group of plants with a clonal relationship, where all plants are from one rootstock, rhizome, stolon, or root system.

Corm: A bulblike underground structure in which the fleshy portion is predominantly stem tissue and is covered by membranous scales.

Corolla: The part of the flower made up of petals.

Corymb: Short, broad, more or less flat-topped, indeterminate inflorescence, the outer flowers opening first.

Decumbent: Reclining or lying on the ground, but with the end ascending.

Determinate: Of limited or definite growth or size; an inflorescence in which the terminal flower matures first.

Dicotyledons: Plants in one of the two subgroups of the angiosperms; characterized by two cotyledons in the embryo.

Dioecious: Having the male and female reproductive organs on separate plants.

Disc: The tubular flowers in the Aster family.

Divided: Cut 3/4 or more the distance from the leaf margin to midrib, or petal apex to base, or, generally, any deep cut.

Drupe: A fleshy, usually single-seeded indehiscent fruit with the seed enclosed in a stony endocarp.

Endemic: Restricted to a relatively small area or region.

Entire: A margin without teeth, lobes, or divisions.

Ephemeral: Lasting only a short time.

Epiphytic: A plant growing upon another plant, but not as a parasite.

Fertile: A flower, or flower part, bearing functional reproductive structures.

Filiform: Threadlike, slender, and usually round in cross section.

Flora: A collective term to refer to all of the plants of an area; a book dealing with the plants of an area.

Flower: An aggregation of highly modified fertile and sterile leaves making up the characteristic reproductive structure of angiosperms.

Flower, complete: A flower in which all four parts (calyx, corolla, stamens, and pistil) are present.

Flower, incomplete: Flowers in which one or more whorl or group of parts is missing.

Flower, perfect: A flower that contains both male and female reproductive organs; it need not have sepals and/or petals and may thus be incomplete.

Follicle: A dry fruit, from a single ovulary, that splits open along a single line.

Fruit: A matured ovulary with or without accessory structures.

Glabrous: Without trichomes or hairs.

Glandular: Having or bearing secreting organs, glands, or trichomes.

Glaucous: Covered with a thin, whitish, waxy "bloom" that can be wiped off.

Herb: A plant with no persistent woody stem above ground.

Herbaceous: Plant parts with little or no hard, woody (secondary) tissue.

Herbarium: A collection of pressed, dried plants, and supporting information, filed by family, genus, and species for use in research and teaching.

Hirsute: With rather rough or coarse trichomes, or hairs.

Indeterminate: Not of limited growth or size; not determinate.

Inflorescence: The flowering portion of a plant.

Involucre: A whorl or collection of bracts surrounding or subtending a flower cluster or a single flower.

Keel: The line of fusion of the two wing petals in legume flowers.

Labiate: With lips, as the bilabiate corolla of many mints.

Lanceolate: Lance-shaped, much longer than wide, widened at or above the base and tapering to the apex.

Leaf: The flattened, usually green, vegetative organ consisting of a distal blade (the flattened part) and a stalk or petiole.

Leaf, compound: A leaf in which the blade is subdivided into two or more leaflets; compound leaves may be pinnate, bipinnate, or palmate.

Leaf, simple: A leaf with only one blade; not compound.

Leaflet: A single unit or division of a compound leaf, which will ultimately separate from the leaf axis by an abscission layer.

Legume: A simple, dry, dehiscent fruit splitting along two sutures; characteristic of the Fabaceae or Bean family.

Linear: Long and narrow with essentially parallel margins, as the blades of most grasses.

Lip: The upper or lower portion of a two-lipped, or zygomorphic, flower.

Lobed: Cut from 1/8 to 1/4 the distance from the margin to midrib, or apex to base, or, more generally, any cut resulting in rounded segments.

Monocarpic: Plant that lives for multiple years, flowers once, and dies.

Monocotyledon: Plants in one of the two subgroups of the angiosperms; characterized by one cotyledon in the embryo.

Monoecious: Having both kinds of incomplete (unisexual) flowers borne on a single plant.

Mycorrhiza: The mutualistic association between certain fungi and the roots of flowering plants.

Node: The point on a stem at which one or more leaves are produced.

Oblique: Slanting; unequal-sided.

Obovate: Inversely ovate.

Opposite leaf arrangement: Two leaves at a single node.

Ovulary: The fertile part of the pistil, enclosing the ovules, often called ovary.

Ovule: The structure occurring inside the ovulary that contains the egg cells or female gamete; after fertilization it becomes the seed.

Panicle: An indeterminate, branching inflorescence.

Pappus: The modified calyx lobes in Asteraceae.

Parallel venation: Leaf venation in which the major veins (vascular bundles) are essentially parallel with one another; relatively characteristic of monocotyledons.

Parasite: A plant that gets its food from another living organism.

Parted: Cut from 1/2 to 3/4 the distance from the margin to midrib, or apex to base, or, more generally, any moderately deep cut.

Pedicel: The stalk of an individual flower in an inflorescence.

Peduncle: The stalk of a flower cluster or of a single flower if the inflorescence consists of a single flower.

Peltate: With the petiole joining the blade near the center rather than at the margin.

Perennial: Plant of three or more years duration.

Perfoliate: A sessile leaf or bract whose base completely surrounds the stem, the latter seemingly passing through the leaf.

Perianth: The calyx and corolla of a flower.

Petal: One of the leaflike appendages making up the corolla.

Petiole: The stemlike part of the leaf.

Pistil: The female reproductive parts of a flower; the stigma, style, and ovulary collectively.

Pistillate: A flower with one or more pistils but no stamens; a female flower, flower part, or plant.

Plicate: Folded, as a paper fan.

Pollination: The transfer of pollen grains from the anther of the stamen to the stigma of the pistil.

Prostrate: A general term for lying flat on the ground.

Pubescence: A general term for hairs or trichomes.

Pubescent: Covered with short, soft trichomes.

Punctate: With translucent or colored dots, depressions, or pits.

Raceme: A simple, elongated, indeterminate inflorescence with pedicelled or stalked flowers.

Ray: A single branch of an umbel; the strap-shaped (ligulate) flowers in the inflorescence of the Aster family.

Reflexed: Abruptly recurved or bent downward or backward.

Rhizome: An underground stem, usually horizontally oriented and sometimes specialized for food storage.

Rosette: An arrangement of leaves radiating from a crown or center and usually at or close to the earth.

Sagittate: Like an arrowhead in form; triangular, with the basal lobes pointing downward or inward toward the petiole.

Saprophyte: A plant that gets its food from dead organic material.

Scabrous: Rough; feeling rough or gritty to the touch.

Scape: A leafless or naked flowering stem.

Scapose: Producing a scape.

Sepals: The outermost, sterile, leaflike parts of a complete flower.

Serrate: With sharp teeth pointing forward.

Sessile: Without petiole or pedicel.

Silique: The long, slender fruit of some species in the Mustard family.

Spatulate: spoon-shaped.

Spur: A tubular or saclike projection from a petal or sepal; also, a very short branch with compact leaf arrangement.

Staminate: A flower with stamens but no pistil; a male flower or plant.

Standard: The upper, usually enlarged, petal of a legume flower.

Stellate: Starlike, with radiating branches.

Stigma: The pollen-receptive, terminal part of the pistil.

Stipule: The basal, paired, leaflike appendages of a petiole, sometimes fused.

Style: The elongated, sterile portion of the pistil between the stigma and the ovulary.

Subtending: Below or beneath, as the bracts subtending an inflorescence.

Succulent: Juicy, fleshy.

Suture: A seam, or a line of opening.

Taxonomy: A branch of botany that deals with the classification and identification of plants.

Tendril: A slender twining appendage or axis that enables plants to climb.

Ternate: In three; three-parted or -divided, as some leaves.

Tomentose: Densely wooly or pubescent, with matted, soft, wool-like hairs.

Trichome: A plant hair; trichomes may be simple, stellate, or glandular.

Tuber: A fleshy, enlarged portion of a rhizome or stolon with only vestigial scales; true tubers are found in the Solanaceae.

Tubular: Having the form of a tube.

Umbel: An inflorescence with pedicels or peduncles (rays), or both, each arising from a common point.

Urceolate: Urn-shaped, as the corollas of some Ericaceae.

Villous: With long, soft, shaggy hairs.

Whorl: Three or more leaves or flowers at one node; in a circle.

Whorled leaf arrangement: Three or more leaves attached to a stem at a single node.

Wing: A thin membranous extension; the lateral petals in Fabaceae and Polygalaceae.

Zygomorphic: A bilaterally symmetrical flower, divisible into equal halves in one plane only, usually along an anterior-posterior line; not actinomorphic, or radially symmetrical.

Wild Flowers *of* North Carolina

Typhaceae

CATTAIL

Typha latifolia Linnaeus

This rhizomatous perennial, 1–2.5 m tall, has flat leaves, 5–24 mm wide. The brown, compact, cylindrical "cattail," which is 0.5–2 dm long, is made up of hundreds of minute female flowers that have each produced a single plumed fruit with a single seed. The many small male flowers form a slender, ephemeral spike above the female flowers and shed copious pollen.

Found in shallow waters of lakes, ponds, and rivers, this common aquatic ranges throughout the state and much of North America. Three other species of Cattail are found in North Carolina. All are easily grown from pieces of the rhizome for bog or water gardens. In colonial times, cattail pollen was collected and added to meal for bread.

MAY–JULY (19-1-1)
A-NB/SGE/INM
[◯/A/$/✪]

Alismataceae

DUCK POTATO

Sagittaria lancifolia Linnaeus
[*Sagittaria falcata* Pursh]

Although other members of this genus have, as the name implies, sagittate, or arrow-shaped, leaves, the narrow leaves of this native perennial are lanceolate, up to 3.5 dm long, and the flowering stem is 6–13 dm tall. Flowers of all *Sagittaria* species are in whorls along the stem, the lower flowers pistillate and the upper flowers staminate.

Native to the tidal marshes, ditches, and stream margins of the outer coastal plain of North Carolina, Duck Potato ranges throughout the Southeast and can be easily grown from seed or root divisions. The underground stems contain a considerable quantity of starch and are eaten by ducks and other animals.

JUNE–OCTOBER (27-3-8)
A-3B/SNE/RRW
[◯/A/$]

Alismataceae

ARROWHEAD; SWAMP POTATO

Sagittaria latifolia Willdenow

These robust native perennials are 6–12 dm tall and bear 2 to 10 whorls of flowers that may be 5 cm broad. The sagittate, arrow-shaped leaves are up to 2.5 dm long and wide, the petioles up to 10 dm long.

Infrequent in bogs, wet ditches, and stream margins, chiefly of the mountains and upper piedmont, Arrowhead ranges broadly throughout North America. The nutritious, starch-filled tubers are a valuable food for waterfowl and were once roasted and eaten by both Native Americans and early settlers as a primary starch source and as a medicinal.

JUNE–SEPTEMBER (27-3-9)
A-3B/SCE/RRW
[○/A/$/✪]

Poaceae

RIVER OATS

Chasmanthium latifolium (Michx.) Yates
[*Uniola latifolia* Michaux]

This rhizomatous native perennial, an attractive "inland cousin" of Sea Oats, has cauline, lanceolate leaves up to 2.5 dm long. These branching plants, up to 1.5 m tall, form large clumps and have loose panicles of dangling, flattened, light green fruits, or "oats," that mature to tan or bronze in the fall.

Occurring in moist woods and river bottoms, River Oats range throughout North Carolina and the eastern and central United States. Easily grown from seed or rhizome cuttings, these plants are recommended for wetland mitigation and stream restoration as well as for the wild garden.

JUNE–OCTOBER (29-10-3)
H-NA/SNE/PNG
[◗/M/$]

Poaceae

SEA OATS

Uniola paniculata Linnaeus

A signature plant of southeastern coastal dunes, this native, rhizomatous perennial grass helps stabilize the shifting dunes, as drifting sand settles in its clumps of leaves. In summer and fall, the graceful slender stalks, each 1–2 m tall, bear many flattened, yellow-brown or straw-colored spikelets.

Sea Oats are characteristic of the beach dunes and adjacent low open areas of the outer coastal plain from Virginia to Texas. These grasses are frequently planted on the dunes and, because of their ecological importance, are protected in North Carolina.

JUNE–JULY (29-10-4)
H-NB/SGE/PNG
[❍/D/$]

Poaceae

WILD RICE

Zizania aquatica Linnaeus

These profusely branched native annuals, 2–3 m tall, have large stem leaves up to 12 dm long and up to 5 cm wide. The yellow to reddish-brown anthers of the many male flowers are conspicuous when in bloom. The long, slender, starchy black grains produced by these plants are similar to other economically important grains in the Grass family and are considered a delicacy.

Important as mast for wildlife, colonies of Wild Rice are found in brackish and freshwater marshes throughout much of eastern North America. Wild Rice is a traditional Native American crop in the northern United States and Canada, where grains are harvested into canoes.

MAY (29-65-1)
A-NA/SLE/PNG
[❍/A/$]

Poaceae

SANDSPUR

Cenchrus longispinus (Hackel) Fernald

Though few people have ever noticed the small flowers on the 2–5 dm long stems of this clumped annual grass, many are painfully familiar with the small (up to 2.5 mm long), spine-covered fruits, or burs. The burs of the sprawling *C. tribuloides* on the dunes of North Carolina's outer coastal plain are larger, 3.5–4 mm long—watch out!

The most widespread of our sand-spurs, in sandy woods and fields, chiefly of the coastal plain, and scattered in the piedmont and mountains, *C. longispinus* ranges throughout much of the United States.

JUNE–OCTOBER (29-69-5)
H-NA/SLE/RNG
[❍/D]

Poaceae

PLUME GRASS; BEARD GRASS

Erianthus giganteus (Walt.) Muhlenberg

This coarse perennial, up to 4 m tall, has many small brownish-red flowers in a colorful plume-like panicle, 1.5–4 dm long and 8–15 cm broad. Four other species of Plume Grass are also native to North Carolina.

This large Plume Grass grows on savannas, in moist ditches, and along woodland margins of the coastal plain and piedmont. Its general distribution is in the South Atlantic and Gulf Coast area.

SEPTEMBER–OCTOBER (29-85-5)
H-NA/SLE/PNM
[❍/M]

Poaceae

BROOM SEDGE

Andropogon virginicus L.

These perennial grasses (not sedges), 5–15 dm tall, have multiple stems from a short rhizome. Leaves are both basal and cauline, up to 3 dm long. Racemes usually are in pairs

As the name implies, these ubiquitous eastern and central North American weeds of roadsides and abandoned fields were once tied in bundles and used as hand brooms. These plants are extremely hardy and add both color and texture to our winter landscape—or wild garden.

SEPTEMBER–OCTOBER (29-87-7)
H-NB/SLE/RNG
[❍/D/✪]

Cyperaceae

WHITE-BRACTED SEDGE

Rhynchospora latifolia (Baldw. ex Ell.) Thomas
[*Dichromena latifolia* Baldwin]

Although the small flowers of sedges are relatively inconspicuous, the seven or more bright white bracts just below the inflorescence of this 3–7 dm tall, grasslike, rhizomatous perennial furnish a colorful contrast to the green of its other leaves and the surrounding low vegetation.

White-bracted Sedge is native to the savannas and wet ditches of our coastal plain and ranges throughout the Southeast to Texas. The similar *R. colorata*, with fewer, narrower bracts, is also found in our area. Both can be maintained in the bog garden without undue effort.

MAY–SEPTEMBER (30-4-1)
H-NB/SLE/KNW
[❍/W/$]

Cyperaceae

COTTON GRASS

Eriophorum virginicum Linnaeus

Despite the common name, this is a sedge and not a grass. This 7–12 dm tall, rhizomatous perennial has both basal and cauline grasslike leaves. When in fruit these plants produce many long, soft, cottonlike bristles in a compact tawny mass, 2–5 cm across, which give them their common name.

An uncommon plant of our mountain or coastal plain bogs, Cotton Grass is more abundant north and west to Manitoba but occurs south to Georgia. Cotton Grass is recommended for the bog garden. ✗

JULY–SEPTEMBER (30-9-1)
H-NB/SGE/KNG
[◐/W/R]

Araceae

GOLDEN CLUB

Orontium aquaticum Linnaeus

This rhizomatous, perennial aquatic, 2–6 dm tall with ovate leaves, 1–4 dm long, is, as for all Araceae, characterized by its spadix of tiny embedded flowers. Unlike the spadix of most other members of the Arum family, the long, slender spadix of Golden Club is not enclosed by a spathe. The fertile portion of the spadix is the golden yellow terminal 5–10 cm, which bears many small greenish-yellow flowers.

Though chiefly found in bogs and along the margins of slow, acid streams of the coastal plain, these native perennials are also found at scattered localities in the piedmont and mountains. Golden Club ranges north to New York and throughout the southeastern states to Texas. It makes a colorful accent (and conversation piece!) in the bog garden. Note that, as with many members of the Arum family, all parts of the plant will cause burning and swelling of the mouth due to the presence of calcium oxylate crystals.

MARCH–APRIL (32-2-1)
A-NB/SOE/INY
[◐/A/$/⚠]

Araceae

SKUNK CABBAGE

Symplocarpus foetidus (L.) Nuttall

The unusual-looking "flower" of this conspicuous perennial is actually a large, somewhat fleshy, reddish-brown spathe, 8–15 cm long, that, as its name implies, emits a strong skunklike odor. The color, odor, and slightly elevated air temperature inside the spathe attract flies, beetles, and other pollinators. The large odorous leaves, eventually 4–5 dm long, appear with, or shortly after, the spathes and at first are rolled into compact cones.

Skunk Cabbage, infrequent in bogs of our mountains and northern piedmont, is more common throughout northeastern North America. It makes an interesting addition to the bog garden. Throughout its range, the roots and leaves of Skunk Cabbage have been used by Native Americans as a remedy for a wide spectrum of ailments.

FEBRUARY–MARCH (32-3-1)
H-NB/SCE/INM
[◗/w/$/⊕/◬]

Araceae

ARROWLEAF; TUCKAHOE

Peltandra virginica (L.) Kunth

The large glossy leaves of this robust, 2–6 dm tall native perennial vary from 1 to 4.5 dm in length. The green spathe, 12–18 cm long, and later the greenish-black fruits are characteristic of this common eastern species.

Found in bogs, marshes, and wet ditches in scattered localities throughout the state, Arrowleaf ranges throughout eastern North America. These plants are easy to maintain in a bog garden—but give them plenty of room for their large leaves.

MAY–JUNE (32-4-1)
A-NB/SCE/TNG
[❍/A/$/✪]

Araceae

WHITE ARROW ARUM

Peltandra sagittaefolia (Michx.) Morong

This showy native perennial, 1.5–5 dm tall, is distinguished from *P. virginica* by its bright white, scapose spathe, 6–12 cm long, red fruits, and smaller leaves, up to 2.5 dm long.

Found at its northern limit in open bogs in only three of our coastal plain counties, this species is more common further south to Florida and Mississippi. With a little winter protection, it is an ideal plant for the bog garden, and is easily grown from seed.

JULY–AUGUST (32-4-2)
A-NB/SCE/TNW
[❍/A/$]

Araceae

GREEN DRAGON

Arisaema dracontium (L.) Schott

The large, solitary, somewhat pinnately cut or divided leaf of this native perennial usually has 7 to 15 segments (vs. always 3 for Jack-in-the-Pulpit) and stands 0.5–1 m tall, well above the rather inconspicuous "flower." The slender, greenish-white spadix is 8–12 cm long and exceeds the spathe.

Infrequent in low woods at scattered localities throughout the state and the eastern United States, this plant makes an interesting but not very showy plant for the bog garden.

MAY (32-5-1)
H-NB/SOE/TNG
[◗/M/$/✪/△]

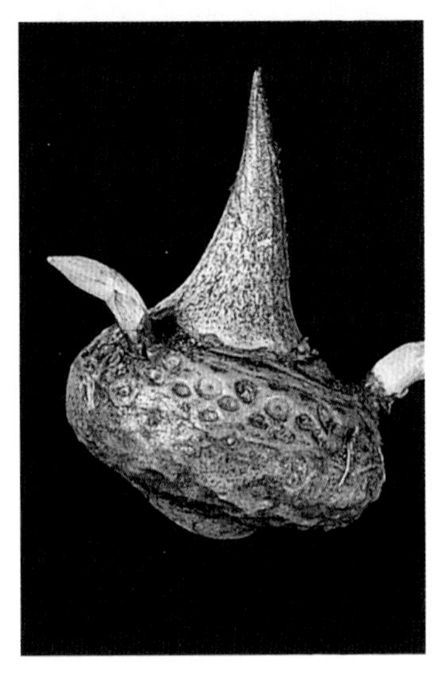

Araceae

JACK-IN-THE-PULPIT

Arisaema triphyllum (L.) Schott

The colorful green spathe of this cormose native perennial, 2–8 dm tall, is often variously striped with maroon and forms the 5–10 cm long "pulpit" more or less surrounding the erect, cylindrical spadix with its minute male and female flowers. The spathe withers in the fall, exposing the cluster of fleshy, scarlet berries.

These interesting plants grow in low woods and bogs at scattered localities throughout the state and in much of eastern North America and can be easily grown for the woodland garden from seeds or corms. The dormant corms can be dug up, potted, and brought inside to force for winter bloom. The raw, starch-filled corm contains calcium oxalate and is very pungent but, as Native Americans knew, is edible when boiled. Native Americans also used the root and other parts of the plant as a medicinal for a wide range of ailments including colds, sores, and snakebite.

MARCH–APRIL (32-5-2)
H-NB/TEE/TNG
[●/M/$/✪/△]

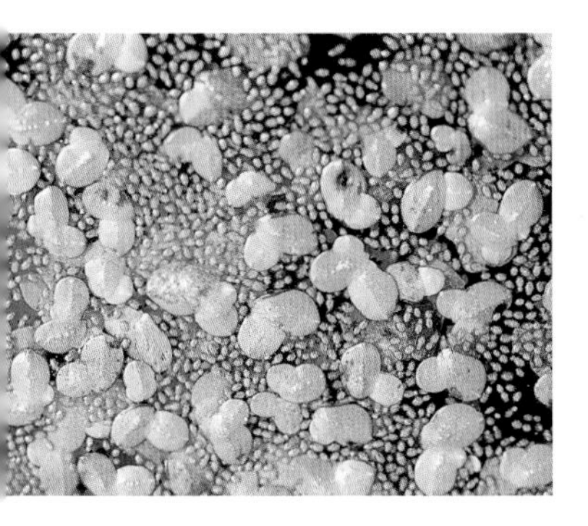

Lemnaceae

DUCKWEED

WATER MEAL

Lemna perpusilla Torrey
Wolffia columbiana Karsten

The small, flat plants of Duckweed and the still smaller plants of Water Meal, both shown here at about natural size, are among the smallest flowering plants in the world. Each flower is reduced to only a single stamen or a single pistil; there are no sepals or petals.

Both Duckweed and Water Meal are found floating on the surface of ponds, marshes, swamps, rivers, and lakes on our coastal plain and range throughout the eastern and central United States south to South America. Water Meal, however, is quite rare in North Carolina. Both plants are more frequent southward and rarely flower in our area.

(33-2-2) (33-3-1)
A-NN/—/SNG
[❍/A/$]

Xyridaceae

YELLOW-EYED GRASS

Xyris fimbriata Elliott

These slender native perennials with grasslike leaves have distinctive yellow flowers produced from compact brown heads, 1–2.5 cm long, at the top of the 5–12 dm tall leafless stems. Only 1 or 2 of the ephemeral flowers, often lasting for only a few hours, open on the headlike spike each day.

Although plants of this particular species are infrequent in bogs, pond margins, and wet ditches in scattered localities in our coastal plain, there are a dozen other species of similar appearance and in similar habitats throughout the state. Indeed, species of Yellow-eyed grass are found throughout eastern North America.

SEPTEMBER–OCTOBER (35-1-1)
A-3B/SGE/IRY
[❍/W]

Eriocaulaceae

PIPEWORT; HATPINS

Eriocaulon compressum Lamarck

The compact heads of these native perennials, made up of many small flowers, are 1–1.5 cm broad and are borne at the top of a slender stem 2–8 dm tall. The linear, parallel-veined leaves, arranged in a rosette at the base of the flower stalk, are generally similar in all four of our species of *Eriocaulon*. The dried heads remain white and are attractive in dried flower arrangements.

Infrequent at scattered localities, chiefly of the coastal plain, Hatpins range variably along the east coast of North America and are quite appropriate for the bog garden.

JUNE–OCTOBER (36-1-3)
H-NB/SGE/KNW
[○/w]

Commelinaceae

DAYFLOWER

Commelina communis Linnaeus

The lower petal of these delicate flowers is white and greatly reduced; thus the corolla appears to be made up of only 2 blue petals. A flower lasts only a day or less, but other buds, within each green spathe (up to 2.5 cm long and seen here just below the flower) continue to open for a week or more. The 2–8 dm long stems of these annuals introduced from China root from the lower nodes and may be erect or sprawling.

Dayflowers are weeds of low, moist, often disturbed areas or waste places at scattered localities in the state and much of the eastern and central United States.

MAY–OCTOBER (38-1-3)
H-3A/SOE/TZB
[○/M/✪]

Commelinaceae

PINK SPIDERWORT

Callisia graminea (Small) G. Tucker [*Tradescantia rosea* var. *graminea* (Small) Anderson & Woodson]

The individual flowers of the Spiderworts, like those of the Dayflowers, last only a few hours. The two differ, among other ways, in that Spiderworts have no spathe. The narrow, weakly clustered leaves of this 2–5 dm tall native perennial are up to 3 dm long but less than 3 mm wide, which differentiates this species from the closely related *C. rosea*, with leaves greater than 3 mm wide.

Plants of this species, which grows from Virginia to Florida, form grasslike clumps in the sandy, open woodlands of the coastal plain of North Carolina and could make a great addition to the rock garden.

MAY–JULY (38-3-1)
H-3B/SGE/URR
[○/D/$]

Commelinaceae

SPIDERWORT

Tradescantia subaspera Ker

A handsome, rather robust, fleshy native perennial, 3–8 dm tall, Spiderwort often forms large clumps when growing under ideal soil and light conditions. The flowers, in which all three petals develop equally, are up to 3 cm across and vary from dark to light blue or even pinkish-blue.

This species is infrequent in dry woods and clearings and along forest margins in North Carolina, chiefly in our mountains and piedmont, but it ranges throughout the eastern United States north to New York and west to Missouri. This species should do well in dry, well-drained soil; the rarer *T. virginiana* is prized for its ease of cultivation in the garden. The leaves of all *Tradescantia* species can cause minor skin irritation.

JUNE–JULY (38-3-2)
H-3A/SNE/URB
[○/D/$/☞]

Pontederiaceae

PICKERELWEED

Pontederia cordata Linnaeus

This somewhat rank native, aquatic perennial, 4–10 dm tall, produces a 5–15 cm long spike of odorous, ephemeral, tubular flowers and glossy, usually wide, long-petiolate leaves, 7–18 cm long.

Relatively frequent in the shallow water of lakes, ponds, streams, and ditches, primarily of the piedmont and coastal plain, Pickerelweed is found throughout eastern and central North America. Although easy to grow (and easy to pull out), this aggressive, rhizomatous aquatic will quickly fill in a pond.

MAY–SEPTEMBER (39-2-1)
A-3B/SCE/IZB
[○/A/$/✪]

Liliaceae

CARRION FLOWER

Smilax herbacea Linnaeus

These green-stemmed native perennial vines, 1–3 m long, have umbellate clusters, 2–5 cm in diameter, of small green flowers. The flowers are easily overlooked, but the strong carrion odor that attracts carrion fly pollinators cannot be missed.

These rhizomatous vines, found in swamp forests and wooded coves of our mountains and piedmont, range through eastern and central North America south to Georgia and Louisiana. The attractive globose fruit clusters are found in the fall. Roots and leaves were used in medicine by several Native American tribes.

MAY–JUNE (41-2-2)
V-3A/SOE/URG
[◗/M/✪]

Liliaceae

LAUREL GREENBRIAR; CATBRIAR

Smilax laurifolia Linnaeus

The thick, glossy, laurel-like, evergreen leaves of this native perennial are 6–10 cm long. These high-climbing, tough, woody, spiny vines often form dense thickets that may grow over and cover other vegetation.

Chiefly a coastal plain plant, it is found frequently in bogs, moist lowlands, and low sandy fill areas behind beach dunes throughout the southeastern United States; eight other evergreen Greenbriar species may be found in a variety of woodland habitats throughout North Carolina. *Smilax* species are easily propagated from sections of the tough woody rhizome. The tuberous roots of Laurel Greenbriar have been used as a starch source and for medicine.

JULY–AUGUST (41-2-10)
V-3A/SEE/URG
[❍/M/✪]

Liliaceae

LITTLE SWEET BETSY

Trillium cuneatum Rafinesque

The single flower of this 1–3 dm tall, native perennial is sessile, or stalkless; the 3 large, ovate leaves, mottled with a lighter green, are 6–18 cm long and wide. The erect petals of "Wet Dog Trillium" (another common name in reference to the flower's scent) are 4.5–6.5 cm long. The berry is maroon red appearing in late May to June.

Infrequent on rich, wooded slopes on circumneutral to basic soils in our southern piedmont and mountains, Little Sweet Betsy can be found over much of the Southeast. The berries and fruits are mildly toxic if ingested. All Trilliums make wonderful additions to the wild garden, but please propagate them by seed; do not dig! ✗

MARCH–APRIL (41-3-1)
H-3W/SOE/SRM
[●/M/$]

Liliaceae

YELLOW TRILLIUM

Trillium luteum (Muhl.) Harbison
[*Trillium viride* var. *luteum* Beck]

These plants are similar to *T. cuneatum* in all respects except for flower color. Although some plants of this species in areas outside North Carolina have green petals, all of our plants are of the yellow variety.

This rare, rhizomatous native perennial is found in deciduous woods on limestone or basic soils in only three mountain counties and has a slightly more restricted range in the Southeast than does Little Sweet Betsy, but it reaches south to Georgia and Mississippi and west to Kentucky and Tennessee. ✗

MARCH–APRIL (41-3-2)
H-3W/SOE/SRY
[●/M/$/*R*]

Liliaceae

MOTTLED TRILLIUM

Trillium discolor Wray

The yellow, 2–5.4 cm long petals of the sessile flower of this 1–3 dm tall Trillium are wider at the tip than at the base and are in sharp contrast to the purple stamens. The leaves, 6–7.5 cm long, are mottled with both light green and purple.

This rare native perennial is known from deciduous woods in only two of our mountain counties and three counties in South Carolina and adjacent Georgia. This species is threatened according to its state rarity status. Preserve! ✗

MARCH–MAY (41-3-3)
H-3W/SOE/SRY
[●/M/*T*]

Liliaceae

NODDING TRILLIUM

Trillium cernuum Linnaeus

The flower stalk, or pedicel, at the tip of the 3–5 dm tall stem of the Nodding Trillium is bent downward; thus the solitary flower is often hidden under the whorl of wide, spreading leaves. Later the 7.5–15 cm long leaves fold slightly upward and the three-sided lavender to purple berry is more exposed.

This attractive perennial is found in rich woods at scattered localities in our mountains and piedmont and ranges throughout northeastern North America, reaching its southern limit in the mountains of Georgia and Alabama. ✗

APRIL–MAY (41-3-5)
H-3W/SOE/SRW
[●/M/$]

Liliaceae

WAKE ROBIN

Trillium erectum Linnaeus

Although sometimes known as Red (or Purple) Trillium, there is also a yellow and a white form of this strongly pedicellate, 2–5 dm tall species that has petals 2–4.5 cm long. The flowers of the dark-colored form are ill-scented and are pollinated by flies that are attracted by their maroon color and carrion odor. The red to purple berry appears in July–August.

These native perennials are chiefly northeastern in general distribution, ranging from Canada through the southern Appalachians in rich, moist woods. Wake Robin is one of the more common Trilliums propagated and available in the trade for the wild garden. Native Americans used this plant in poultices for tumors and inflammation as well as for other ailments. ✗

APRIL–JUNE (41-3-6)
H-3W/SOE/SRM
[●/M/$/✪]

Liliaceae

SWEET WAKE ROBIN

Trillium vaseyi Harbison

This plant is sometimes considered only a variety of *Trillium erectum*, which it resembles except for the larger petals, 3.5–6 cm long and often overlapping, and its habitually nodding flowers.

These native perennials are infrequent in moist woods of their limited range in the mountains of the southern portion of the Blue Ridge in North Carolina, South Carolina, Georgia, and Tennessee. This Trillium is readily available in the trade for the wild garden. ✗

APRIL–JUNE (41-3-6b)
H-3W/SEE/SRM
[●/M/$]

Liliaceae

CATESBY'S TRILLIUM

Trillium catesbaei Elliott

The 1.5–4 dm tall stems of this pedicellate Trillium have flowers that are about 4 cm across and may be nodding or on a level with the short-petiolate leaves.

Probably our most abundant Trillium, this native rhizomatous perennial of the southern Appalachians is usually found in the deciduous woods and forests of the lower mountains and piedmont of North Carolina. It is found in only four other states: South Carolina, Georgia, Alabama, and Tennessee. ✗

APRIL–JUNE (41-3-7)
H-3W/SEE/SRR
[●/M/$]

Liliaceae

LARGE-FLOWERED TRILLIUM

Trillium grandiflorum (Michx.) Salisbury

The flower of this striking and variable native perennial is on a short stalk or pedicel. The firm white petals may be up to 5 cm long and turn pink as they age after pollination. The plants are 2–5 dm tall.

This Trillium is restricted to rich woods, usually on basic soils, in our mountains, where it may be found at widely scattered localities, often in spectacular colonies, and northward into Canada. This plant was used for medicine by the Cherokee and Iroquois throughout its range. ✗

APRIL–MAY (41-3-8)
H-3W/SOE/SRW
[●/M/$/✪]

Liliaceae

PAINTED TRILLIUM

Trillium undulatum Willdenow

The white flowers, with their distinctive red inverted "V" at the base of each petal, make plants of this perennial especially attractive and easy to identify. Stems are 1–4.5 dm tall, and the ovate leaves are 8–16 cm long.

Although more common northward into Canada, these plants occur in the scattered surviving bogs and moist hemlock or spruce-fir forests of North Carolina's mountain counties. They reach their southern limits in Georgia. The scarlet berry can be seen in July–August. This is a highly popular Trillium for the wild garden. ✗

APRIL–MAY (41-3-10)
H-3W/SOE/SRW
[●/M/$/✪]

Liliaceae

INDIAN CUCUMBER ROOT

Medeola virginiana Linnaeus

The two whorls of leaves and the nodding flowers, about 1.5 cm in diameter, help identify this 2–8 dm tall native perennial in bloom; the colorful change in the upper leaves and the dark blue berries with red pedicels identify it in the fall.

Found in moist, deciduous woods throughout North Carolina and eastern North America, Indian Cucumber is more frequent in the mountains and piedmont of our state. As suggested by the name, the root tastes like cucumber and was used as food and medicine by Native Americans.

APRIL–JUNE (41-4-1)
H-3W/SEE/URY
[●/M/$/✪]

Liliaceae

CLINTON'S LILY

Clintonia borealis (Ait.) Rafinesque

The leathery leaves of *Clintonia*, 1–4 dm long, have a strongly depressed midvein. The 4 to 8 cream-colored or yellow flowers of this 1.5–6 dm tall, native perennial have petals 1.5 cm or more long, and form a loose cluster near the top of the slender scape, or flower stalk; the berries are clear blue.

Clinton's Lily is near its southern limits in the few remaining spruce-fir forests and on the heath balds at high elevations in our mountains but is more abundant further north, ranging throughout northeastern North America. Native Americans made extensive use of the leaves and roots to treat infections and to aid labor in childbirth, and the root has a compound with both anti-inflammatory and estrogenic activity of potential use in modern medicine. ✗

MAY–JUNE (41-5-1)
H-3B/SEE/UFY
[●/M/$/✪/R]

Liliaceae

SPECKLED WOOD LILY

Clintonia umbellulata (Michx.) Morong

This 2–3.5 dm tall perennial is very similar to *C. borealis* but differs by having smaller, white flowers, petals 1 cm long, a more compact flower cluster, and ciliate, or hairy, leaf margins. The fruits are dark blue to black.

Speckled Wood Lily is found at scattered localities in the wooded coves of our mountain counties and ranges in the mountains from New York to Georgia. ✗

MAY–JUNE (41-5-2)
H-3B/SEE/UFW
[●/M/$/✪]

Liliaceae

FALSE SOLOMON'S SEAL

Maianthemum racemosum (L.) Link
[*Smilacina racemosa* (L.) Desfontaines]

The 7 to 13 leaves of this 2–10 dm tall native perennial are 3–8 cm wide, and the flower cluster, or inflorescence, is 10 cm or more long. The stem is conspicuously angled in a zigzag fashion from one set of leaves to the next. The dull red berries ripen in August–October.

These rhizomatous plants are scattered in deciduous woodlands of eastern and central North America and are found throughout North Carolina except on the outer coastal plain. Try this attractive Lily in the wild garden. Native American tribes throughout the range of False Solomon's Seal used the roots, leaves, and berries to treat a wide variety of ailments, and several tribes used the berries and roots for food or forage.

APRIL–JUNE (41-6-1)
H-3A/SEE/PRW
[●/M/$/✪]

Liliaceae

FALSE LILY-OF-THE-VALLEY; CANADA MAYFLOWER

Maianthemum canadense (L.) Desfontaines

The fragrant flowers of this low plant, 4–15 cm tall, are quite different in shape from those of the true Lily-of-the-Valley, but the one or two dark green leaves, 3–10 cm long and 2–5 cm wide, do somewhat resemble those of *Convallaria*. If undisturbed, these native, rhizomatous perennials can form extensive colonies.

This delicate Lily is infrequent in the dwindling spruce-fir forests and wooded coves of our mountains but is a bit more common northward, ranging through northern North America west to Montana and south to Georgia. This interesting little plant has a history of both Native American and folk medicinal use.

MAY–JUNE (41-7-1)
H-3A/SOE/RRW
[●/M/$/✪]

Liliaceae

NODDING MANDARIN

Prosartes maculata (Buckl.) A. Gray [*Disporum maculatum* (Buckl.) Britton]

These rare perennials, 1–8 dm tall, are freely branching, the branches usually forked. The creamy white petals, 2–2.5 cm long and 5–8 mm wide, are spotted with purple. The related Yellow Mandarin, *P. lanuginosa*, has yellow petals that are smaller and not spotted.

Found in rich woods in only two counties in North Carolina's mountains, Nodding Mandarin is more common north of our state but ranges from southern Michigan as far south as Georgia and Alabama. Yellow Mandarin is found north to southern Ontario, ranging through the mountains to Georgia. ✗

APRIL–MAY (41-8-2)
H-3A/SEE/URW
[●/M/$/*R*]

Liliaceae

TWISTED STALK

Streptopus lanceolatus var. *roseus* (Michx.) Reveal
[*Streptopus roseus* Michaux]

These 3–8 dm tall native perennials produce one or two small, rose-colored flowers, 0.6–1 cm long, from each axil of the upper leaves. The slender pedicels, or flower stalks, have an offset, or bend, near the middle, which accounts for the common name. The red berries ripen in July.

Infrequently found in the few remaining spruce-fir forests and wooded coves of our mountains, where it is at its southern limit, Twisted Stalk is more frequent northward, where it ranges into Canada. Native Americans used this plant for both food and medicine. ✗

APRIL–JUNE (41-9-1)
H-3A/SEE/SFR
[●/M/$/✪]

Liliaceae

SOLOMON'S SEAL

Polygonatum biflorum (Walt.) Elliott

The long, arching stems of this relatively common native perennial vary from 0.2 to 2 m in length. It is easily distinguished from False Solomon's Seal by its larger flowers, and later by its blue-black fruits, borne in small axillary clusters along the stem rather than in a terminal inflorescence.

Plants of this widespread species of eastern and central North America are found scattered in deciduous woods and clearings generally throughout North Carolina and are ideal for the woodland garden. Both Native American and folk medicine used a tea made from the roots of Solomon's Seal. The berries are mildly toxic when ingested.

APRIL–MAY (41-10-2)
H-3A/SNE/SFW
[●/M/$/✪/△]

Liliaceae

AMERICAN LILY-OF-THE-VALLEY

Convallaria majuscula Greene
[*Convallaria montana* Rafinesque]

Two or three dark green leaves, 1.5–3 dm long and 5–13 cm wide, and small, white, bell-shaped, very fragrant flowers, on a scape 1–2 dm long, characterize this low perennial. Unlike its commonly cultivated European relative, this American cousin does not form large colonies.

Found in rich woods at scattered localities in North Carolina, this Lily-of-the-Valley ranges in the Appalachians from Pennsylvania to Georgia. All parts of these plants are toxic when ingested in large amounts.

APRIL–JUNE (41-11-1)
H-3B/SEE/RCW
[●/M/$/△]

Liliaceae

BEAR-GRASS; YUCCA

Yucca filamentosa Linnaeus

The leathery, linear, 2–6 dm long, evergreen leaves of this native perennial have slender filaments, or strands of fibers, along the margins. The stout flower stalk is 1–3 m tall.

This conspicuous perennial is often cultivated. Native along the Atlantic Coast from New Jersey to Florida and escaped further inland, it is found in old fields and open woodlands at scattered localities throughout North Carolina. Another, and larger, species of Yucca, *Y. aloifolia*, is especially plentiful in the low sandy areas behind the beach dunes of our outer coastal plain. The Cherokee used Yucca to stupefy fish and as a soap, as well as for medicine.

MAY–JUNE (41-12-3)
H-3B/SLE/PFW
[❍/D/$/✪]

Liliaceae

TURKEY BEARD

Xerophyllum asphodeloides (L.) Nuttall

Turkey Beard's grasslike clumps of basal leaves, up to 4 dm long and 2 mm wide, are easily overlooked, and its spectacular flower stalks, 0.8–1.5 m tall, are frequently chewed off by deer. Thus this rare plant is even more rarely noticed or photographed.

In our area Turkey Beard grows in dry, open woods, chiefly in the mountains. It generally ranges in the Appalachian region and a portion of the coastal plain from New Jersey to Virginia. ✗

MAY–JUNE (41-15-1)
H-3B/SGE/RRW
[◐/D/*R*]

Liliaceae

SWAMP PINK

Helonias bullata Linnaeus

The stout, hollow, flowering stem of this native perennial, 1–3 dm tall in flower and up to 6 dm in fruit, has many small, scalelike leaves or bracts. The larger evergreen, spatulate leaves, 1–2 dm long, form a basal rosette. The attractive inflorescence, 3–8 cm long, is at the top of a 1–3 dm long scape.

A very rare plant of mountain swamps and bogs known from only four of our counties, the species has a general range along the mountains from Georgia to New York. This beautiful Lily is of special concern, with both a federal and state threatened status, and should be actively preserved. Be sure any plants offered for sale are indeed nursery grown. Do not dig or disturb! ✗

APRIL–MAY (41-16-1)
H-3B/SBE/RRR
[◐/W/$/*T-SC*]

Liliaceae

BLAZING STAR; DEVIL'S BIT; FAIRYWAND

Chamaelirium luteum (L.) Gray

These plants are dioecious, that is, one plant bears only female or pistillate flowers and another bears only male or staminate flowers. Male plants of Blazing Star, shown here, are 3–7 dm tall and are far more attractive than the taller (up to 12 dm) female plants. Male plants are thus the choice for the wild garden.

These native perennials are found in rich, deciduous woods at scattered localities throughout the state and the eastern United States. In herbal medicine, the staminate plants were used for men's ailments and the pistillate plants for female problems.

MARCH–MAY (41-17-1)
H-3A/SEE/RRW
[●/M/$/✪]

Liliaceae

RUSH-FEATHERLING; PLEEA

Pleea tenuifolia Michaux

The slender flowering stems of this fall-blooming native perennial are 3–8 dm tall and about twice as long as the narrow, mostly basal grasslike leaves. The delicate flowers have separate petals, 10–15 mm long, and are quickly followed by ovate capsules.

These plants are infrequent in moist, open, sandy areas of a few counties of the southeastern coastal plain of North Carolina, where they may be found in association with Venus Flytrap. Their general range extends south to Florida.

SEPTEMBER–OCTOBER (41-19-1)
H-3B/SGE/RRW
[○/W]

Liliaceae

WHITE COLICROOT; STARGRASS

Aletris farinosa Linnaeus

The 4–12 dm flower stalk of this native perennial grows from the center of a basal rosette of widely linear to lanceolate leaves, 8–20 cm long. The small flowers, 6–8 mm long, have a granular or mealy appearance when observed closely and can add texture in a dried arrangement—or the wild garden.

Relatively frequent on the savannas and moist, sandy roadsides of the coastal plain, more scattered in the moist woods and meadows of the piedmont and mountains, this species ranges more or less throughout the eastern United States. As the common name implies, the roots of White Colicroot have been used as a digestive tonic.

MAY–JULY (41-20-2)
H-3B/SNE/RTW
[❍/N/$/✪]

Liliaceae

FLYPOISON

Amianthium muscaetoxicum (Walt.) Gray

The dense raceme of this native perennial is 3–13 cm long on a slender stalk 2.5–8 dm tall. The mostly basal leaves are 1–6 dm long and 4–23 mm wide. The white petals and sepals do not wither after the flower has been pollinated but persist on the plant and turn green as they age.

A relatively widespread southeastern perennial ranging north to New York and west to Oklahoma, Flypoison is found throughout North Carolina in bogs, savannas, and moist, open deciduous woodlands. As the name implies, the plant is highly poisonous, especially the bulb; it was used by the Cherokee as a crow poison.

MAY–JULY (41-21-1)
H-3B/SLE/RRW
[❍/M/✪/◉]

Liliaceae

DEATHCAMUS

Zigadenus leimanthoides Gray

The open, branching inflorescence immediately separates this 4–7.5 dm tall, poisonous, native perennial from Flypoison. The linear leaves, 3–6 dm long and 4–12 mm wide, are gradually reduced upward. There are four species of *Zigadenus* in North Carolina, and Crow Poison (*Z. densus*), which is common on the coastal plain, does closely resemble Flypoison.

Native to the southeastern states, these plants are quite rare in North Carolina. Known only from heath balds in two of our mountain counties, deathcamus is state listed as significantly rare. All parts of these plants are highly toxic. ✗

JULY–AUGUST (41-22-4)
H-3A/SLE/PRW
[❍/M/◉/*SR*]

Liliaceae

BUNCH-FLOWER

Melanthium latifolium Desrousseaux [*Melanthium hybridum* Walter]

An attractive native perennial, Bunch-Flower has a flowering stalk 1–2 m tall; the linear leaves, 2–8 dm long and 0.4–3 cm wide, are both basal and cauline. The branched, terminal panicle of racemes is 2–5 dm long.

These plants are frequent in deciduous forests at scattered localities in the mountains and piedmont of North Carolina and range through the Atlantic Coast states north to New York.

JULY–AUGUST (41-23-2)
H-3A/SLE/PRW
[❍/M/$]

Liliaceae

FALSE HELLEBORE

Veratrum viride Aiton

This coarse native perennial has a single leafy stem 6–15 dm tall bearing numerous yellow-green flowers on the short upper branches. The wide, ovate leaves, 1.5–3 dm long, end in a sharp point and are strongly ribbed or plaited along the more or less parallel veins.

A northern species that is relatively frequent and conspicuous in bogs and moist woodlands of our mountain counties, False Hellebore has a long history of medical use and is currently used in pharmaceutical drugs to slow heart rate and lower blood pressure. All parts of this plant are highly toxic, the roots being especially potent.

JUNE–AUGUST (41-24-1)
H-3A/SOE/PCG
[●/M/$/✪/◉]

Liliaceae

FEATHERBELLS

Stenanthium gramineum (Ker) Morong

The single, tall, slender flowering stalk of these perennials, 3–15 dm tall, ends in a branched inflorescence of numerous, small white to green flowers with linear sepals and petals, 4–8 mm long. The flowers on the lateral branches are mostly staminate, while those in the main or terminal spike are perfect.

Frequent in the mountains in moist meadows, bogs, and deciduous forests at scattered localities in the piedmont and coastal plain, Featherbells range throughout the Southeast as far west as Texas and as far north as Pennsylvania and Michigan.

JULY–SEPTEMBER (41-25-1)
H-3A/SLE/PRW
[❍/M]

Liliaceae

TROUT LILY; DOGTOOTH VIOLET

Erythronium americanum Ker

Trout Lilies are among our earliest spring flowers. The wide, mottled, basal leaves, 0.5–2 dm long, and the 1–2 dm scape with nodding yellow flowers with strongly recurved sepals and petals immediately identify these native perennials. Look carefully for the obovoid, green capsule that lies on the ground as it matures and dehisces to release the small, ant-dispersed seeds.

Often found in large, ancient, slow-growing colonies in open deciduous woodlands of our mountains and piedmont, and rarely in the coastal plain, Trout Lily occurs throughout much of eastern North America south to Georgia and Alabama. Used by Native Americans for medicine, water extracts from the plant have been shown to have antibacterial activity. Dogtooth Violet can be propagated by seed and is recommended for the moist woodland garden, but it is usually four years from planting to flower. Good luck!

FEBRUARY–APRIL (41-26-1)
H-3B/SEE/SFY
[●/M/$/✪]

Liliaceae

PINE LILY

Lilium catesbaei Walter

This attractive coastal perennial, 3–10 dm tall, is easily identified by the (usually) single, erect flower at the tip of the alternate-leaved stem. The linear leaves are 2–8 cm long and 2–15 mm wide.

Pine Lily grows in savannas, bogs, and pine flatwoods on the coastal plain from southeast Virginia to Louisiana and makes a wonderful addition to the wild flower garden, but do not dig plants from the wild. ✗

JUNE–SEPTEMBER (41-32-2)
H-3A/SLE/SFY
[❍/M/$]

Liliaceae

GRAY'S LILY

Lilium grayi S. Watson

The single stem of these perennials is 6–12 dm tall and has several whorls of 4–8 lanceolate, 4–13 cm long, leaves. There are 1 to 4 horizontal or slightly nodding flowers, each on a long stalk. The flowers, about 4–5 cm long, are not strongly flared or reflexed.

This extremely rare native plant is found in high meadows and balds of the mountains of North Carolina, Tennessee, and Virginia. If you are fortunate enough to acquire viable seeds, or plants properly propagated by seed, Gray's Lily is a prize for the wild garden. These plants are federally and state listed; do not dig! ✗

JUNE–JULY (41-32-3)
H-3W/SNE/UFR
[❍/M/*T-SC*]

Liliaceae

CANADA LILY; WILD YELLOW LILY

Lilium canadense Linnaeus

The nodding orange-yellow flowers with flared sepals and petals, 5–8 cm long, distinguish this 6–15 dm tall native perennial from *L. grayi*. The whorled leaves, 4 to 12 at each node, are 8–15 cm long.

Canada Lily is rare, with a state listing, in wet meadows and woodlands in two of North Carolina's counties, but it is more common northward, ranging from Quebec and Maine south to Georgia and Alabama. Native Americans made use of the roots for both medicine and food. ✗

JUNE–JULY (41-32-4)
H-3W/SNE/UFY
[○/M/✪/*SR*]

Liliaceae

TURK'S CAP LILY

Lilium superbum Linnaeus

Plants of this showy perennial may be 3 m tall and bear a dozen or more nodding flowers (rarely up to 100 or more in fertile soil in full sun). The strongly reflexed sepals and petals are green at the base, giving the flower a dark center. The acutely pointed lanceolate leaves, 5–20 per node, are 8–18 cm long and 1–3 cm wide.

Found in moist or wet meadows and coves of our mountains, and along the Blue Ridge Parkway, Turk's Cap Lily ranges from New Brunswick to Minnesota and south to Florida and Mississippi. This Lily is readily available from native plant nurseries; do not dig it from the wild. ✗

JULY–AUGUST (41-32-5)
H-3W/SNE/UFY
[◗/N/$]

Liliaceae

MICHAUX'S LILY; CAROLINA LILY

Lilium michauxii Poiret

Although plants of this species may appear similar to the Turk's Cap Lily, they are smaller, 0.4–1.3 m, and have only 1 to 6 flowers. Also, the petals lack the green basal markings of the Turk's Cap and the fleshy, obtusely pointed leaves are widest above the middle.

These uncommon but relatively widespread southeastern perennials are found more or less throughout North Carolina in upland pine-oak woods and pocosins, ranging more generally from Virginia to Texas. In 2003 the state legislature named the Carolina Lily the state wild flower. (Dogwood is still our state flower.) ✗

JULY–AUGUST (41-32-6)
H-3W/SBE/UFY
[◗/N/$]

Liliaceae

LARGE-FLOWERED BELLWORT

Uvularia grandiflora Smith

The large, drooping flowers of this 2–7.5 dm tall native perennial are 3–5 cm long and are often almost hidden by the rolled or curled upper leaves, which are 6–14 cm long.

These plants are scattered in forests on limestone or calcareous soils over much of eastern and central North America south in the mountains to Georgia. In North Carolina there are three other interesting and attractive species of *Uvularia* that are ideal for the wild flower garden. *U. grandiflora* was used to make root tea, and two others, *U. perfoliata* and *U. sessilifolia*, were eaten as greens and used for medicine by Native Americans.

APRIL–MAY (41-33-2)
H-3A/SEE/SFY
[●/M/$/✪]

Liliaceae

NODDING ONION

Allium cernuum Roth ex Roemer

The scape, or flower stalk, of this native perennial, with the distinct odor of onion, is 2–6 dm tall. The flat leaves are glaucous, or whitish, 1–4 dm long and only 3–8 mm wide.

Common locally in open meadows and deciduous woods and on granite outcrops of our mountains and piedmont, this species has a general range over much of North America where suitable habitats exist. A good choice for the rock garden, Nodding Onion is easily grown from seed or bulb division and has a long history of Native American use for both medicine and food.

JULY–SEPTEMBER (41-35-5)
H-3B/SGE/URW
[○/N/$/✪]

Liliaceae

RAMPS

Allium tricoccum Aiton

This bulbous, perennial spring ephemeral is most readily identified by the two flat elliptic-lanceolate leaves and its strong scent. The leaves appear in early spring and disappear by the time the lone flower stalk, 1.5–3 dm tall, bears a single umbellate inflorescence with pale yellow, campanulate flowers followed by three-lobed capsules.

Ramps are distributed broadly in northeastern and north central North America south in the mountains to North Carolina, where they can be locally abundant in rich cove woodlands. They have become quite popular as a spring vegetable, collected early in the season before flowering. Ramps, along with other species of *Allium*, have also been used for folk and Native American medicine. Ramp festivals are held in many places throughout the mountains and are contributing to overcollection and extirpation of these interesting (and pungent!) plants from our dwindling forests.

JUNE–JULY (41-35-7)
H-3B/SEE/URW
[●/M/$/✪]

Dioscoreaceae

WILD YAM

Dioscorea villosa L.

These perennial, dioecious, twining vines, up to 3 m long, have small greenish, inconspicuous flowers and are easily overlooked. However they can often be spotted by the whorl of thin, prominently veined, ovate, acuminate, deciduous leaves, 5–10 cm long, with a strongly cordate base on the lower part of the vine. The leaves are whorled lower and alternate toward the tip of the vine. Unlike the closely related Cinnamon Vine, Wild Yam has no small, asexual, potatolike tubers in the axils.

Wild Yam is common in woodlands throughout the state and throughout eastern and central North America. A tea made from its roots was used by Native Americans for a variety of ailments, and, along with other members of the genus *Dioscorea*, it has important uses in modern medicine.

APRIL–JUNE (43-1-2)
V-3W/SCE/PCG
[●/N/✪]

Amaryllidaceae

SPIDER LILY

Hymenocallis crassifolia Herbert

These smooth, scapose, bulbous perennials, 3–5 dm tall, are easily identified by their large, distinctive flowers, with linear petals and sepals 5–10 cm long, in 1 to 3 flowered umbels. The flowers are made even showier by the white "crown" of thin tissue that connects the bases of the anthers.

Spider Lily reaches its northern limit in ditches, along rivers, and in brackish marshes in our southern coastal plain and extends south in the coastal plain to Florida. The bulbs of all species of *Hymenocallis* have a low toxicity when ingested. ✗

MAY–JUNE (44-3-1)
H-3B/SLE/URW
[○/W/$/△/*R*]

Amaryllidaceae

ATAMASCO LILY

Zephyranthes atamasco (L.) Herbert

These low perennials, 1–2.5 dm tall, have narrow, linear leaves, 2–4 dm long, and erect, funnel-form flowers 7–10 cm long that rapidly change from pure white to pink following pollination.

Native to the southeast from Virginia to Louisiana, Atamasco Lily is usually found in wet meadows and low woods in the coastal plain and piedmont of North Carolina but is also present in a few western counties. The mountain town of Cullowhee takes its name from the Indian name for this Lily.

The Atamasco Lily is a wonderful choice for the wild garden, but all parts of this species, especially the bulb, are highly toxic, even deadly, when eaten.

MARCH–APRIL (44-4-1)
H-3B/SLE/SFW
[❍/M/$/◉]

Amaryllidaceae

YELLOW STARGRASS

Hypoxis hirsuta (L.) Coville

These low, grasslike native perennials have pubescent leaves, 1–6 dm long. The shorter flower stalk (up to 4 dm at maturity) bears a loose cluster of 1 to 3 buds that open over a period of several days; the yellow flowers are 1–2.5 cm across.

A widespread species of eastern and central North America, these plants are frequently found in old wagon ruts through open woodlands and meadows throughout North Carolina. The plants may be either solitary or scattered at a given locality and are an excellent choice for the rock garden or wild garden.

MARCH–JUNE (44-7-1)
H-3B/SGE/URY
[◗/D/$/✪]

Amaryllidaceae

AGAVE; FALSE ALOE; RATTLESNAKE MASTER

Manfreda virginica (L.) Salisbury ex Rose [*Agave virginica* Linnaeus]

Unlike their large western relatives, the monocarpic "Century Plants" (that live and grow for many years, then bloom once and die), the eastern Agave is a relatively small perennial, up to 2 m tall, with a whorl of evergreen, succulent basal leaves, 1.5–4 dm long. The flower stalk produces an open raceme of mottled greenish-white flowers, 2.5–3 cm long, with a distinctive spicy fragrance that attracts dusk-flying, moth pollinators.

Agave inhabits open woodlands on basic soils in our piedmont and southwest mountain areas and ranges across the mideastern and central United States. As the name Rattlesnake Master suggests, the plant has indeed been used to treat snakebites among other medicinal uses. Easy to grow, Agave is a top choice for the rock garden—or just to plant in a corner of your yard for its fragrance.

MAY–JULY (44-8-1)
H-3B/SNE/RTG
[◗/D/$/✪]

Haemodoraceae

REDROOT

Lachnanthes caroliniana (Lam.) Dandy

These 3–9 dm tall perennials have flat, linear, basal leaves that are 0.5–1.5 cm wide and 2–5 dm long. The compact, branched, tomentose inflorescence expands somewhat as the flowers mature. The red sap of the roots and rhizomes accounts for the common name.

Common in sandy low savannas, ditches, and pocosin borders of our coastal plain, these impressive plants extend along the coast from Massachusetts to Louisiana.

JUNE–SEPTEMBER (45-1-1)
H-3B/SLE/URW
[❍/w]

Iridaceae

BLUE-EYED GRASS

Sisyrinchium angustifolium Miller

In full sun, these native perennials often form large clumps that produce many flower stalks. The 1.5–5 dm long, flattened or winged flowering stems of these grasslike plants are much like the somewhat shorter leaves in appearance. Despite the small size of an individual flower, 8–12 mm wide, the clumps, which are easily divided, are quite colorful.

A frequent plant in meadows and open, moist woodlands, these plants are found essentially throughout the state and much of eastern and central North America. Several Blue-eyed Grasses have been used medicinally and are suitable for the rock or wild garden.

MARCH–JUNE (46-2-4)
H-3B/SGE/SRB
[○/N/$/✪]

Iridaceae

BLUE FLAG

Iris virginica L.

This attractive native perennial is 5–10 dm tall, with a flowering stalk that may branch, bearing 2 to 6 flowers up to 8 cm in diameter. Leaves (flattened into a single plane as for all Iris species) are 4–11 dm long and 1–3 cm wide, and sometimes top the flowering stalk.

Although rare in the piedmont, Blue Flag is found in marshes, swamps, and stream margins throughout North Carolina and the Southeast from Virginia to Texas. An obligate wetland plant, Blue Flag tolerates partial shade, is easy to grow, and makes an ideal plant for wet garden habitats. Blue Flag has documented use by Native Americans as a medicinal, but note that *all* Iris species have roots that are poisonous when ingested.

APRIL–MAY (46-5-2)
H-3B/SLE/STB
[◗/W/$/✪/⚠]

Iridaceae

DWARF IRIS

Iris verna Linnaeus

The flowering stems of this Iris and the next are usually only 2–5 cm long, much shorter than our other native and cultivated *Iris* species. The orange band on the uncrested sepals, the distinct fragrance, and the relatively narrow, straight leaves distinguish this species from the Crested Iris.

This rhizomatous perennial is infrequent in open, often rocky woods at scattered localities throughout the state; it is native from New York to Florida west to Arkansas. This species is a great choice for the wild garden that grows well in full sun or broken shade. Cover the clumps with an unobtrusive dome of poultry wire (held in place with ground staples) to keep the deer from eating them—flowers, leaves, and rhizome!

MARCH–MAY (46-5-7)
H-3B/SLE/STB
[◗/D/$/✪/△]

Iridaceae

CRESTED DWARF IRIS

Iris cristata Aiton

Although the wide leaves of these 2–7 cm tall plants may grow up to 4 dm in length, they are relatively short when the plants are in bloom. The sepals have a ciliate or pubescent crest in a white area near the base, which gives this plant its name and separates it from *I. verna.*

This infrequent, rhizomatous, native perennial of rich woods of our mountains and piedmont ranges from Pennsylvania to Georgia west to Oklahoma. Both Dwarf Iris species make colorful additions to the wild garden, but protect them from hungry deer.

APRIL–MAY (46-5-8)
H-3B/SLE/STB
[◗/N/$/✪/△]

Orchidaceae

MOCCASIN FLOWER; PINK LADY'S SLIPPER

Cypripedium acaule Aiton

The inflated lower petal of this wild orchid forms a moccasinlike pouch 3–5 cm long. The leafless flowering stem, or scape, is 2–4 dm tall and the two basal leaves, 1–2 dm long, are silvery beneath and green above.

These native perennials, which range throughout northeastern North America south through the mountains to Georgia, often form large colonies in low pinelands and bogs, chiefly of our mountains and coastal plain. Like other orchids, they will grow only when certain symbiotic fungi, or molds, are present in their roots. If this fungus cannot survive in an area, the orchid will not live there either; transplants always die when their fungal partner dies. Some tissue-cultured and seed-propagated Lady's Slippers are available in the trade. The leaves of all Lady's Slipper species have glandular hairs that may cause minor skin irritation. Do not dig! ✗

APRIL–JULY (49-1-1)
H-3B/SEE/SZR
[◗/M/$/✪/☞/*RM*]

Orchidaceae

YELLOW LADY'S SLIPPER

Cypripedium pubescens Willdenow
[*Cypripedium calceolus* var. *pubescens* (Willd.) Correll]

These showy native perennials, 2–8 dm tall, may be either solitary or in spectacular colonies that take years to form (do not disturb!). The 3 to 5 alternate leaves on the flowering stem are 3–10 cm wide. The lip, or lower petal of the flower, is inflated to form a pouch 2–6 cm long.

Found on rich wooded slopes in the mountains and a few scattered localities in the piedmont, this Lady's Slipper is circumboreal, ranging south through much of the United States but is, nonetheless, quite rare as a result of years of overcollection for medicinal use. Several different varieties of this species occur throughout the eastern states, but none is available for the garden. Tissue culture is the only possibility for the propagation of these beautiful but "difficult" orchids. ✗

APRIL–JUNE (49-1-2)
H-3A/SEE/SZY
[●/M/$/✪/☞/*RM*]

Orchidaceae

SHOWY LADY'S SLIPPER

Cypripedium reginae Walter

The rarest of the three species of Lady Slipper occurring in North Carolina, these coarse, pubescent perennials are 4–10 dm tall and, like *C. pubescens*, have leafy flower stalks. The large, ovate leaves, 5–10 cm long, are strongly ribbed; the flowers are fragrant.

A more northern species than the others, native in swamps and on cool, wooded slopes in two of our mountain counties, this species ranges throughout northeastern North America south through the mountains to Georgia and west to Arkansas. ✗

MAY–JUNE (49-1-3)
H-3A/SEE/SZR
[◗/M/$/☞/*RM*]

Orchidaceae

SHOWY ORCHIS

Galearis spectabilis (L.) Rafinesque
[*Orchis spectabilis* Linnaeus]

The Showy Orchis's short scape, arising between the two lustrous, 3–10 cm wide, basal leaves, is usually 1–2 dm tall and bears 3 to 10 colorful flowers. The sepals and two petals are magenta, while the third (lower) petal forms the flat white lip.

A rare perennial, with a range from New Brunswick to Minnesota south to Georgia and Arkansas, Showy Orchis grows along streams in rich hardwood (or rarely pine) forests of our piedmont and mountains. As with other wild orchids, it cannot be moved. Do not dig! ✗

APRIL–MAY (49-2-1)
H-3B/SOE/RZR
[●/M/*RM*]

Orchidaceae

GREEN FRINGED ORCHID

Platanthera lacera (Michx.) G. Don
[*Habenaria lacera* (Michx.) Loddiges]

The lacerated or finely cut lip of these small, fragrant flowers gives them the fringed appearance responsible for their name. The flower stalk is 2.5–7.5 dm, tall with several oblong lower leaves, 7–21 cm long, and reduced, bractlike upper leaves.

An infrequent to rare native perennial in bogs, marshes, and wet meadows of eastern North America south to Georgia, in North Carolina it occurs chiefly in the mountains but is also known from several stations in the piedmont and coastal plain. ✗

JUNE–AUGUST (49-3-1)
H-3A/SLE/RZG
[❍/W/*RM*]

Orchidaceae

SMALL PURPLE FRINGED ORCHID

Platanthera psycodes (L.) Lindl.
[*Habenaria psycodes* (L.) Sprengel]

These native, somewhat variable perennials, from 1.5 to 9.5 dm tall, have elliptic leaves, 5–22 cm long, on the lower stem. The numerous individual flowers of the showy, compact raceme are 3–20 cm long with a 3-lobed, ciliate lip, 4–9 mm long.

Infrequent but conspicuous in open woods, seepages, and upland meadows of the North Carolina mountains, this orchid ranges through northeastern North America south through the mountains to Georgia. Please leave it for others to enjoy! ✗

JUNE–AUGUST (49-3-3)
H-3A/SEE/RZB
[◗/N/✪/*RM*]

Orchidaceae

PURPLE FRINGELESS ORCHID

Platanthera peramoena (Gray) Gray
[*Habenaria peramoena* Gray]

The 3-parted lip or lower petal of these flowers is 1.5–2 cm across. In contrast to *P. psycodes*, the margins of the 3 segments are not cut so deeply as to appear fringed but are only slightly serrate or toothed. The raceme is 6–18 cm long and 4–6 cm broad on a stalk 3–10 dm tall.

This rare orchid occurs in moist woods and meadows only in our mountains and is found more broadly in the Mid-Atlantic states west to Missouri. Please do not disturb! ✗

JUNE–OCTOBER (49-3-4)
H-3A/SEE/RZB
[◗/M/*RM*]

Orchidaceae

ROUND-LEAVED ORCHID

Platanthera orbiculata (Pursh) Lindley
[*Habenaria orbiculata* (Pursh) Torrey]

The 1–6 dm tall flowering scapes of this attractive native perennial arise between a pair of distinctive flat, rounded basal leaves, 7–25 cm long and 4.5–19 cm wide. The greenish-white flowers have spurs, 1.5–4.5 cm long, and are in racemes of up to 20 or more flowers.

This unique orchid ranges from Labrador to Alaska south in the mountains to our northwestern counties, where it reaches its southern limits in moist coniferous and deciduous forests and bogs. Please do not disturb! ✗

JUNE–SEPTEMBER (49-3-7)
H-3B/SRE/RZW
[◗/M/✪/*RM*]

Orchidaceae

WHITE FRINGED ORCHID

Platanthera blephariglottis (Willd.) Lindley
[*Habenaria blephariglottis* (Willd.) Hooker]

The compact spike of this attractive orchid is 5–15 cm long atop a 4–8 dm tall stalk. The slender lip of the relatively small flower is finely fringed and contrasts with the entire, spreading sepals.

This rare to infrequent native perennial is found in bogs and moist, peaty depressions in savannas and pine woodlands of our coastal plain and one mountain county. White Fringed Orchid ranges from Newfoundland and Ontario south to Virginia and along the coastal plain to Florida and Mississippi. Do not dig! ✗

JULY–SEPTEMBER (49-3-13)
H-3A/SLE/RZW
[◗/M/*RM*]

Orchidaceae

YELLOW FRINGED ORCHID

Platanthera ciliaris (L.) Lindley
[*Habenaria ciliaris* (L.) R. Brown]

This slender native perennial, 2.5–10 dm tall, has linear to lanceolate lower leaves and a few reduced, alternate, stem leaves. The flowers, in spikes 5–15 cm long, each have a long spur and a deeply fringed, or ciliate, lip.

Infrequent and often solitary in bogs, meadows, and along the margins of thickets in our mountains and on our coastal plain, this species ranges from New England to the Carolinas and Georgia west to Texas. Please do not dig or cut! ✗

JULY–SEPTEMBER (49-3-14)
H-3A/SBE/RZY
[◯/M/✪/*RM*]

Orchidaceae

KIDNEYLEAF TWAYBLADE

Listera smallii Wiegand

The scientific name of this tiny orchid, only 1–3 dm tall, does not refer to its size but indicates that the plant was named for the botanist, J. K. Small, who discovered it. The two ovate leaves are 2–4 cm long or broad.

A rare and inconspicuous perennial of the Appalachian Mountains from Pennsylvania to Georgia, Twayblade is found in bogs, in the humus of Rhododendron thickets, or on moist, wooded slopes in western North Carolina. ✗

JUNE–JULY (49-4-1)
H-30/SOE/RZG
[●/M/RM]

Orchidaceae

THREE BIRDS ORCHID

Triphora trianthophora (Sw.) Rydberg

The 1–3 dm stem of this rather delicate native perennial usually bears three flowers, each 1.5–2 cm long, with spreading lateral sepals—the three birds. The alternate, widely ovate leaves, 8–20 cm long, clasp the stem.

Rare but apparently widely scattered in suitable habitats over much of eastern and central North America, these plants are found in rich, damp woods and thickets only in our mountain counties. Do not dig! ✗

JULY–SEPTEMBER (49-5-1)
H-3A/SOE/RZR
[●/M/RM]

Orchidaceae

LARGE WHORLED POGONIA

Isotria verticillata (Willd.) Rafinesque

This delicate orchid's very slender flower, with narrow, almost linear sepals, 3.5–6 cm long, arises directly from the center of a whorl of five wide leaves at the end of a stem 1–4 dm tall.

Found in moist hardwood forests and along stream margins throughout the eastern United States, this interesting perennial is infrequent at scattered localities throughout North Carolina and should be photographed, not dug! ✗

APRIL–JULY (49-6-1)
H-3W/SEE/SZG
[●/M/M]

Orchidaceae

ROSE-CRESTED ORCHID

Pogonia ophioglossoides (L.) Ker

The lip formed by the lower petal is 1.5–2.5 cm long and is densely bearded or crested in this orchid. This native perennial, 1–7 dm tall, has a single, lanceolate stem leaf, 2–12 cm long; the basal, lanceolate to oblong leaves are absent at flowering.

Generally ranging over much of eastern and central North America at lower elevations, this orchid is found occasionally in North Carolina in open bogs and on seepage slopes of the coastal plain and piedmont, and rarely in similar habitats in the mountains. ✗

MAY–JUNE (49-7-1)
H-3B/SBE/SZR
[❍/M/M]

Orchidaceae

ROSEBUD ORCHID

Cleistes divaricata (L.) Ames

The narrow, spreading brown sepals of this perennial orchid are 3–6.5 cm long and contrast sharply with the two fused magenta-pink upper petals and the darker striped, slightly fringed lip. The stem is 2–6 dm tall with usually a single basal leaf and another single firm or leathery oblong leaf, 3–20 cm, halfway up the stem.

This southeastern native ranges from New Jersey to Florida and Mississippi and is found on the savannas and damp pine barrens of our coastal plain and in the upland woods of our mountains. It does not survive transplant. ✗

MAY–JULY (49-8-1)
H-3A/SEE/SZR
[◯/M/*M*]

Orchidaceae

BOG ROSE

Arethusa bulbosa Linnaeus

This low, native perennial, 1–4 dm tall, grows from a bulblike root. The solitary terminal flower, 2–4.5 cm long, appears before the single linear leaf.

One of our rarest plants, and state listed, Bog Rose grows in acid sphagnum bogs in only three of our mountain counties, but it ranges from Newfoundland west to Minnesota and south to New Jersey and along the mountains to Georgia. Do not disturb! ✗

MAY–JUNE (49-9-1)
H-3A/SLE/SZR
[◗/W/*EM*]

Orchidaceae

GRASS PINK

Calopogon tuberosus (L.) B.S.P.
[*Calopogon pulchellus* (Salisb.) R. Brown]

This plant gets its common name from its linear, grasslike leaves and its 4 cm or more broad pink-magenta-crimson flowers. The slender flowering stem is 3–7 dm tall.

Ranging throughout eastern and central North America, this perennial occurs in fairly large colonies in bogs, moist savannas, meadows, and low grassy roadside ditches, chiefly in our coastal plain and mountains. ✗

APRIL–JULY (49-10-2)
H-3B/SLE/RZR
[◯/W/M]

Orchidaceae

NODDING LADIES' TRESSES

Spiranthes cernua (L.) L. C. Richard

The two rows of this wild orchid's small flowers form a double spiral around the top of the flower stalk, which is 2–4.5 dm tall. The linear to lanceolate basal leaves may be 2 dm or more long.

These widespread and somewhat variable perennials are native to much of eastern and central North America, where they are found in varied moist, open habitats. In North Carolina they occur at scattered localities over the state. They rarely, if ever, survive transplant. ✗

JULY–FROST (49-12-2)
H-3B/SLE/IZW
[◯/M/✪/M]

Orchidaceae

SLENDER LADIES' TRESSES

Spiranthes lacera (Raf.) Rafinesque
[*Spiranthes gracilis* (Bigel.) Beck]

The single row of strongly spiralled flowers shows how this genus got its name. The slender spike is 4–10 cm long on a stem 2–7 dm tall. The leaves of these perennials are usually absent at flowering time. The lip of the small flowers is 4–6 mm long.

Ranging through much of eastern and central North America, this species is found in dry to moist fields, meadows, and sandy hardwood forests of the mountains and piedmont of North Carolina. ✗

AUGUST–SEPTEMBER (49-12-4)
H-3B/SLE/IZW
[◗/N/M]

Orchidaceae

DOWNY RATTLESNAKE PLANTAIN

Goodyera pubescens (Willd.) R. Brown

These native perennial orchids, 2–4 dm tall, are densely pubescent and have distinctive, 3–6 cm long, ovate leaves with white reticulate veins. The compact raceme of small flowers is 4–10 cm long. Another species occurring in North Carolina, *G. repens*, 1–3 dm tall, has a fewer-flowered, looser raceme, 3–6 cm long.

This relatively abundant orchid is found in small, isolated clumps or colonies in dry coniferous or hardwood forests in our mountains and piedmont and throughout much of eastern and central North America. Although often transplanted to terraria, these "cute" little plants rarely, if ever, survive for more than a few weeks; leave them in the woods to live. ✗

JUNE–AUGUST (49-13-1)
H-3B/SOE/RZW
[●/D/✪/M]

Orchidaceae

LILY-LEAVED TWAYBLADE

Liparis lilifolia (L.) L. C. Richard

This showy native perennial, 1–2 dm tall, with two shiny, oval, basal leaves 5–15 cm long, has a scape 1–2 dm tall with 5 to 30 brownish flowers in a single raceme.

Infrequent along stream banks and on moist forest slopes, primarily in our mountains and coastal plain, this orchid ranges from New England west to Minnesota and south to Georgia and Arkansas. ✗

MAY–JULY (49-15-1)
H-3B/SOE/RZM
[●/N/M]

Orchidaceae

CRANE-FLY ORCHID

Tipularia discolor (Pursh) Nuttall

The slender flower stalk, 3–5 dm tall, and the small brown and white flowers of this native perennial often escape notice. This is especially true since the characteristic ovate leaves, 5–10 cm long, brownish-green on top and maroon-purple beneath, are absent at flowering time.

This orchid is relatively frequent but inconspicuous in open deciduous woods throughout North Carolina and ranges from New York through the eastern and central United States to Texas. If you must try transplanting orchids, try *Tipularia*; with proper soil, moisture, and care, it might live.

JULY–SEPTEMBER (49-16-1)
H-3B/SOE/RZM
[●/N/*M*]

Orchidaceae

PUTTY ROOT; ADAM AND EVE

Aplectrum hyemale (Muhl.) Torr.

The single, large, elliptic, flat, or weakly pleated leaf, 10–15 cm long, is finely lined with white along the veins and is quite distinctive lying flat against the forest floor in winter. This uncommon perennial produces a 3–6 dm flower stalk in May or June with a raceme of 7 to 15 white or creamy purple flowers.

Putty Root grows in low and rich woods of our mountains and northern piedmont. It ranges over the eastern and central United States but is threatened in much of the Northeast. As with most orchid species, this plant is not a candidate for the wild garden. Do not collect! ✗

MAY–JUNE (49-18-1)
H-3B/SOE/RZW
[●/M/✪/*M*]

Orchidaceae

SPOTTED CORAL ROOT

Corallorhiza maculata Raf.

The succulent, leafless, brownish to purplish flower stalks of these saprophytic orchids may be 1–5 dm tall and bear few to many racemose flowers that vary from pale to dark red. Flowering stalks may occur singly or in a cluster. As with some other orchid species, Spotted Coral Root, although a perennial, does not come up every year, apparently depending on the availability of nutrients from its buried dead wood "host."

Infrequent in our mountains, Spotted Coral Root has a broad distribution in rich soils of upland woods and along stream banks across Canada and the northern United States south in the mountains to the Carolinas. ✗

JULY–AUGUST (49-20-1)
H-3N/—/RZR
[●/M/✪/M]

Saururaceae

LIZARD'S TAIL

Saururus cernuus Linnaeus

The drooping spikes of small fragrant flowers and the pointed, heart-shaped leaves, 5–12 cm long, are characteristic of these 5–9 dm tall, native, rhizomatous, marsh perennials that often form large colonies.

A species of the eastern and central United States, Lizard's Tail is found along stream and lake margins and in swamps and low woodlands of the coastal plain and piedmont. Warning: this plant often becomes invasive and will take over its aquatic environment if not thinned out regularly.

MAY–JULY (50-1-1)
A-NA/SCE/RRW
[◗/A/$/✪]

Betulaceae

CHERRY BIRCH; SWEET BIRCH

Betula lenta Linnaeus

Like those of many of our deciduous trees, the flowers of Alders, Birches, and Oaks are wind pollinated. The flowers are organized into catkins, the male flowers usually in long, drooping catkins that mature before the leaves, releasing enormous quantities of pollen. Female flowers are in smaller, less conspicuous, usually ovoid or conelike catkins with expanded stigmas for the collection of the wind-blown pollen. Cherry Birch is a tree to 25 m tall with a reddish-brown, exfoliating bark and twigs that, when crushed, have the scent of wintergreen. The ovate leaves are 5–10 cm long.

This northeastern species ranges through the mountains to Georgia and is found in our area only in the mountains in rich woods and heath balds at lower elevations. Native Americans used Cherry Birch bark tea to treat fevers and other ailments. The essential oil distilled from the bark was used in folk medicine before being produced synthetically. Warning: the essential oil is toxic and easily absorbed through the skin.

MARCH–APRIL (54-3-2)
T-NA/SOS/INY
[●/N/$/✪/△]

Fagaceae

AMERICAN CHESTNUT

Castanea dentata (Marsh.) Borkhausen

Once the dominant tree of eastern hardwood forests, this majestic, native relative of the Oaks, which often grew up to 30 m tall, has been wiped out by an introduced blight to which it was not immune. Today, a few stump sprouts may grow large enough to bloom and set a few spiny burs, 5–6 cm in diameter, which (if pollinated) may contain 1 to 3 sweet chestnuts.

These bushy stump sprouts can be found occasionally in our mountains and upper piedmont in dry, rich hardwood forests, but without reproduction the species will soon become extinct. Native Americans used the Chestnut as a food source, medicine, fiber, and dye.

JUNE–JULY (55-2-1)
T-NA/SNS/IRG
[●/D/$/✪/*R*]

Fagaceae

CHINQUAPIN

Castanea pumila (L.) Miller

This native, 2–5 m tall shrub or small tree has ellipitic leaves, 6–20 cm long, that are similar to those of its close relative, the American Chestnut. The slender catkins of strong-scented, staminate flowers are 4–15 cm long; the pistillate flowers are inconspicuous. The bur, up to 2.5 cm in diameter, usually contains a single sweet nut.

Scattered in dry, deciduous woods throughout the Southeast north to New Jersey, Chinquapin is found throughout North Carolina. In the late 1960s, when the first edition of this book was written, Chinquapin was thought to be unaffected by the blight that at that time had already wiped out the American Chestnut. However, that is not the case, and our Chinquapins are now seriously affected by the Chestnut blight and will likely suffer the same fate.

JULY (55-2-2)
S-NA/SES/IRG
[◗/D/$/✪]

Urticacaeae

STINGING NETTLE

Urtica dioica Linnaeus

This rhizomatous, clonal perennial, 1–1.3 m tall, has opposite, cordate leaves, 5–10 cm long. All above-ground parts of these plants are covered with stinging hairs. Plants are dioecious or monoecious; the flowers are in axillary panicles, 4–5 cm long. The closely related Wood Nettle, *Laportea canadensis*, also has stinging hairs but has alternate leaves.

This almost cosmopolitan weed is both native and introduced. The typical variety, of European origin, is found in our area in only a few mountain counties. Wood Nettle is abundant in woodlands of our mountains and piedmont. Stinging Nettle has a long history of medicinal use, particularly as an anti-inflammatory for rheumatism and arthritis, and is in current medicinal use in Europe. Native Americans also cooked this plant for greens and for fiber.

MAY–JULY (59-2-1)
H-40/SCS/PRG
[◗/N/✪/☞]

Santalaceae

BASTARD TOADFLAX

Comandra umbellata (L.) Nuttall

These low, herbaceous perennials, 1–4 dm tall, are partly parasitic on the roots of oak trees. The narrowly elliptic leaves are 1–4 cm long, and the small umbels of white flowers are in the leaf axils.

Ranging essentially throughout the United States, *Comandra* is found at scattered localities in the deciduous woodlands of the piedmont of North Carolina and also in a few counties of the mountains and coastal plain. Note: if you transplant a parasite, you must move it with its host or it will die! ✗

APRIL–JUNE (60-1-1)
H-5A/SEE/URW
[●/N/✪/P]

Aristolochiaceae

DUTCHMAN'S PIPE

Aristolochia macrophylla Lamarck

The interesting flowers of this high-climbing perennial woody vine have no petals; the strongly curved sepals form the 2–4 cm long, pipelike bloom. The alternate, heart-shaped leaves are 8–24 cm long and wide.

These impressive vines, primarily of the Appalachian region from New York to Georgia, are infrequent in the rich woods of our mountains. Unfortunately, the older mature stems of these slow growing, woody vines are being seriously overcollected for their flexible woody stems used to make "rustic" hanging baskets.

MAY–JUNE (62-1-1)
V-3A/SCE/STM
[◗/M/$/✪]

Aristolochiaceae

WILD GINGER

Asarum canadense Linnaeus

The thin, deciduous, pubescent heart-shaped leaves of this native, rhizomatous perennial are 7–15 cm wide. Both the leaves and rhizomes give off a gingerlike aroma when broken or crushed. The cup-shaped apetalous flowers, 1–2 cm long, are formed by the fused sepals.

Widespread throughout the eastern half of the United States, *Asarum* is found in rich woods in our mountains and at a few scattered localities in the piedmont. Wild Ginger is a good plant for the cooler-climate wild garden. Use of this species was widespread by eleven Native American tribes throughout its range for many and varied ailments. Wild Ginger was also commonly used to season food.

APRIL–MAY (62-2-1)
H-3B/SCE/STM
[●/M/$/✪]

Aristolochiaceae

HEART LEAF; WILD GINGER

Hexastylis shuttleworthii (Britt. & Baker) Small

The aromatic perennial *Hexastylis* species differ from *Asarum* in their firm, glabrous, evergreen leaves that are often mottled with a lighter green and their fleshy flowers with a wide-lobed calyx. The larger leaves, 6–10 cm long, and flowers, 1.5–4 cm long, differentiate *H. shuttleworthii* from seven other *Hexastylis* species found in our area.

This species occurs in rich woods of the mountains from Virginia to Georgia, but one or more of North Carolina's other evergreen Heart Leaf species can be found in deciduous woods in nearly every county of the state. At least one species, *H. arifolia*, has had medicinal use by Native Americans. With care (shade, water, and good soil), several of our species can be grown in the wild garden.

MAY–JULY (62-3-7)
H-3B/SCE/STM
[●/M/$]

Polygonaceae

SORREL

Rumex hastatulus Elliott

The small flowers of these dioecious winter annuals are crowded on the short branches of the inflorescence at the top of a stem 4–10 dm tall. In all members of the family, the stem has swollen nodes that are sheathed by a ring of stipules called an ocreae.

A common weed of old fields in the coastal plain, often in colorful association with the blue Toadflax, Sorrel is a native of sandy soils at lower elevations over much of the Southeast and further west to New Mexico. Several related weedy species have been introduced, including the similar but smaller, more widespread, rhizomatous perennial weed, Sheep Sorrel (*R. acetosella*) and the larger, robust Sour Dock, *R. crispus*, both used for food and medicine by Native Americans.

MARCH–MAY (63-2-2)
H-6A/SCE/RRR
[❍/N]

Polygonaceae

KNOTWEED; PINKWEED

Polygonum pensylvanicum Linnaeus

The annual, glabrous Pinkweed, one of our most common Knotweeds, has flowers in compact, 1.5–3 cm long, terminal racemes on stalks 0.5–1.5 m tall. Different species of *Polygonum* may be annual or perennial, and the sepals of their apetalous flowers may vary in color from green to white or pink. The stems of all species of Knotweed found in North Carolina are thickened at the nodes and appear jointed. The sheathing ocreae in Pinkweed are not bearded.

This widespread Knotweed is usually found in moist fields, on roadsides, and in disturbed habitats throughout our state and ranges from Canada throughout much of the continental United States.

JULY–FROST (63-4-8)
H-5A/SNE/RRR
[❍/M/✪]

Polygonaceae

TEARTHUMB

Polygonum sagittatum L.

The slender, erect or decumbent, branching, 1–2 m long stems of this native annual are armed with small, sharp, retrorse barbs that can cause shallow skin cuts—thus the common name. Flowers are arranged in compact heads at the end of the flowering stalk. The white (rarely pink) calyx enlarges in fruit up to 2–3.5 mm in length.

This common weed is found in open, moist areas across our state and broadly throughout eastern North America and west to Colorado.

MAY–FROST (63-4-18)
H-5A/SCE/IRW
[○/M]

Polygonaceae

CLIMBING BUCKWHEAT

Polygonum cilinode Michaux

These native, perennial, trailing or twining herbs, up to 2 m long, with smooth stems and alternate, cordate-sagittate leaves, can literally blanket the ground or other vegetation. The numerous small white flowers are in terminal or axillary, paniculate racemes. The ocreae sheathing the stem are distinctively bearded.

These vines thrive in the full sun of forest clearings at higher elevations of our mountains, ranging from eastern Canada through the mountains to Georgia.

JUNE–SEPTEMBER (63-4-20)
H-5A/SCE/PRW
[○/M]

Chenopodiaceae

LAMB'S QUARTERS; PIGWEED

Chenopodium album Linnaeus

This erect, freely branched annual, up to 1 m tall, is covered in a white, mealy pubescence that accounts for its specific name. The 3-nerved rhombic-lanceolate to rhombic-ovate leaves, 3–7 cm wide (the largest leaves irregularly toothed), often turn red late in the season. The tiny greenish flowers lack petals and are tightly clustered in a terminal panicle.

This weedy native, found throughout North Carolina and all of North America, is widely designated a noxious weed. Native Americans used the leaves for a variety of medicinal purposes, and the leaves are still collected, cooked, and eaten as greens—they are not bad!

JUNE–FROST (64-3-2)
H-NA/SND/PRG
[❍/N/✪]

Phytolaccaceae

POKE WEED

Phytolacca americana Linnaeus

These rank, herbaceous, native perennials from 1 to 3 m tall have a spread to match. The large, smooth, lanceolate to elliptic leaves are 8–30 cm long and 3–12 cm wide. The racemes of small white flowers, green fruits, and fleshy, black, ripe fruits (which are deadly poison!) are all present on a large plant at the same time late in the season.

These plants are frequent in low grounds and recent clearings throughout the state and the eastern half of the United States. The berries, roots, and leaves were used in both Native American and folk medicine for rheumatism, arthritis, dysentery, and many other ailments. The juice of the berries was once used for ink. All parts of the plant are poisonous, but the tender leaves are excellent when thoroughly parboiled and properly cooked for "poke salad."

MAY–FROST (68-1-1)
H-5A/SNE/RRW
[❍/N/✪/◉]

Portulacaceae

SPRING BEAUTY

Claytonia virginica Linnaeus

These low, herbaceous perennials, from a globose corm, are 0.5–4.5 dm tall with linear leaves up to 2 dm long. Each corm may produce several to many flowering stems, and each stem can produce up to a dozen flowers with distinctive pink guidelines on the 10–15 mm long petals. The closely related *C. caroliniana* differs only in its elliptic leaves that are 1.5–7 cm long.

These spring ephemerals are found in the low, rich woods of our mountains and piedmont and range through most of the eastern and central United States north to southern Canada. The closely related *C. caroliniana* is restricted to the rich woods of our mountains. *C. virginica* was used for medicine, and the corms of both species were used as a starch source by Native Americans. Plants of both species are essentially "no maintenance" and are ideal for the wild garden.

MARCH–APRIL (70-1-1)
H-50/SLE/RRR
[●/M/$/✪]

Portulacaceae

SEDUM

Talinum teretifolium Pursh

These small, rhizomatous perennials, 1–3.5 dm tall, have fleshy, linear leaves, up to 4 cm long that are round, or terete, in cross section. The petals, 5–10 mm long, drop soon after the flower opens.

Occurring in small pockets of soil on rock outcrops and on dry, sandy soils at scattered localities in the mountains and piedmont of North Carolina, *Talinum* ranges from Pennsylvania to Georgia. This is a good candidate for the rock garden.

JUNE–SEPTEMBER (70-2-1)
H-5A/SLE/URR
[○/D]

Caryophyllaceae

GIANT CHICKWEED

Stellaria pubera Michaux

The attractive flowers of this native, 1–4 dm tall perennial are about 1 cm across and appear to have 10 petals. On closer examination, however, one can see that each flower has only 5 petals, as do other members of the Pink family, but each petal is deeply parted into two linear segments.

Giant Chickweed is relatively frequent in the deciduous woods of our mountains and piedmont, ranging through the eastern United States north to New York. It is an excellent low plant for the wild shade garden.

APRIL–JUNE (71-7-4)
H-50/SEE/URW
[●/M/$]

Caryophyllaceae

SOAPWORT; BOUNCING BET

Saponaria officinalis Linnaeus

Bouncing Bet is a robust, introduced, glabrous perennial, 5–15 dm tall, with pink to white flowers, the petals 8–18 mm long. The opposite, elliptic leaves are 3–10 cm long.

These plants, often cultivated, are widely naturalized throughout the United States and are relatively frequent at scattered localities throughout North Carolina. When well established, they form large, rather showy clumps along roadsides, in old fields, and in waste places. The first common name comes from the fact that the juice of the crushed stems forms a soaplike lather with water. Note: the roots and seeds cause a low toxicity when ingested.

MAY–FROST (71-15-1)
H-50/SEE/UTR
[○/N/$/✪/⚠]

Caryophyllaceae

STARRY CAMPION

Silene stellata (L.) Aiton f.

These native perennials, 5–10 dm tall, have clearly 5-parted flowers with fimbriate petals, 10–20 mm long. The lanceolate stem leaves, each 3–10 cm long, are mostly in whorls of 4.

Though it sometimes occurs in large patches locally, Starry Campion is infrequent in, or at the margins of, rich woods in our mountains and piedmont and ranges through most of the eastern half of the United States. Try it (from seed) in your wild garden.

JULY–SEPTEMBER (71-17-1)
H-5W/SNE/PTW
[◗/M/$]

Caryophyllaceae

FIRE PINK; INDIAN PINK

Silene virginica Linnaeus

The sticky stems of this 2–7.5 dm tall native perennial are regularly branched and have opposite leaves. The crimson petals, 15–25 mm long, are often deeply notched.

Single plants or clumps of Fire Pink are seen along roadsides at the edge of rich woods in the mountains and piedmont of North Carolina and over much of the eastern United States. Fire Pink is a weak perennial but makes a spectacular addition to your native garden if allowed to reseed and maintain a visible presence.

APRIL–JULY (71-17-7)
H-5O/SEE/UTR
[◗/N/$/✪]

Caryophyllaceae

CAMPION

Silene ovata Pursh

This rare native perennial, 3–15 dm tall, has opposite, ovate leaves, 5–12 cm long. The paniculate-cymose branched inflorescence has numerous white flowers with deeply cleft petals, the linear segments 5–10 mm long.

Found in rich woods in only four of our mountain counties, this attractive Campion is known from 11 Appalachian states and is most abundant in the southern part of its range.

AUGUST–SEPTEMBER (71-17-9)
H-50/SOE/PTW
[◗/M]

Nymphaeaceae

COW LILY; SPATTER DOCK

Nuphar advena (Aiton) W. T. Aiton
[*Nuphar luteum* (L.) Sibthorp & Smith]

The floating, suborbicular to lanceolate leaves of these coarse, rhizomatous, easy to grow, aquatic perennials are quite variable and may be up to 5 dm long and 3 dm wide. The fleshy, yellow flower is 2–5 cm across, with numerous stamens and a prominent, fleshy, lobed stigma.

These natives of the eastern United States are found in ponds, rivers, and lakes at scattered localities, chiefly in our coastal plain and piedmont. Cow Lily is easily propagated from sections of the thick rhizome. The roots have been used as a folk remedy for impotence and by Native Americans to treat a variety of diseases.

APRIL–OCTOBER (73-1-1)
A-VA/SCE/SCY
[❍/A/$/✪]

Nymphaeaceae

WATER LILY

Nymphaea odorata Aiton

This native, rhizomatous, aquatic perennial has large, rounded leaves, up to 3 dm wide, that are green above and usually reddish brown beneath. The fragrant white or pink flowers, 8–15 cm across, have numerous 2–10 cm long petals and stay open for several days.

Water Lily is found in North Carolina in lakes, ponds, and wet ditches at scattered localities, chiefly in the coastal plain, but it ranges throughout North America. These plants are ideal for ponds and water gardens. Collect a few rhizomes and start them in your pond—but don't let them take over! This plant has a rich history in both Native American and folk medicine.

JUNE–SEPTEMBER (73-2-1)
A-VA/SCE/SCW
[○/A/$/✪]

Ranunculaceae

YELLOWROOT

Xanthorhiza simplicissima Marshall

A small colonial shrub, 3–5 dm tall, with yellow wood in both the stem and the root (thus the common name), Yellowroot has alternate, pinnately compound leaves. The small apetalous flowers of these interesting plants have brown sepals and are pollinated by equally small insects.

Yellowroot grows along shaded stream banks throughout the southeast (except, apparently, Florida). Roots have been used as a source of yellow dye, and the entire plant has been used as a treatment for stomach ulcers, colds, cancer, and many other ailments in both Native American and folk medicine. Easy to grow, Yellowroot can make an interesting addition to the moist shade garden.

APRIL–MAY (76-1-1)
S-5A/BOS/PRM
[●/M/$/✪]

Ranunculaceae

WILD COLUMBINE

Aquilegia canadensis Linnaeus

The nodding red and yellow flowers of Wild Columbine, 3–4 cm long counting the deep spur at the base of each petal, are pollinated by hummingbirds and are borne on slender stems, 3–12 dm tall. The mostly basal leaves are divided into three rounded segments that may be divided again into three segments.

A native perennial of eastern North America, Columbine is found occasionally in rich, rocky woods at scattered localities throughout North Carolina, but occurs more frequently in the mountains. It is easily propagated from seed and makes a wonderful addition to the wild garden. Native Americans used the roots, leaves, and seeds to treat stomach problems, headache, heart trouble, and many other ailments.

MARCH–MAY (76-2-1)
H-5A/BOE/PTR
[◗/N/$/✪]

Ranunculaceae

LARKSPUR

Delphinium tricorne Michaux

These attractive native perennials, 2–6 dm tall, have colorful and interesting flowers in which the sepals instead of the petals are showy; the single spur, 8–20 mm long, is formed by the upper sepal. The fruit of this species is strongly 3-parted; the leaves are usually divided into 5 primary segments that may be further lobed or cleft.

This Larksur species is infrequent but may occur in large colonies in rich woods of our mountains and lower piedmont and ranges through the Mid-Atlantic states west to Nebraska. Larkspur can be propagated by seed for the wild garden. Note: all Larkspur species are highly poisonous if ingested!

MARCH–MAY (76-3-2)
H-5A/SRC/RZB
[◗/M/$/✪/◉]

Ranunculaceae

MONKSHOOD

Aconitum uncinatum Linnaeus

A weak-stemmed, sprawling, up to 1.5 m tall, native perennial of the Buttercup family, Monkshood has a spurred upper sepal and flowers 15–22 mm long, which are usually blue or purple, but may be white or yellow. The leaves have 3 to 5 lobes that are less dissected than those of *Delphinium*.

Found in rich woods and moist, open meadows from Pennsylvania to northern Georgia, Monkshood is infrequent in our mountains and northern piedmont. It can be propagated by seed, but all parts of the plants are highly poisonous and may prove fatal if eaten.

AUGUST–SEPTEMBER (76-4-1)
H-5A/SOL/SZB
[◗/M/$/◉]

Ranunculaceae

MARSH MARIGOLD

Caltha palustris Linnaeus

No relation at all to the Marigold of our gardens (which belongs to the Aster family), these very rare, showy, native, 2–6 dm tall perennials have 5 to 9 golden, petal-like sepals 8–20 mm long.

This circumboreal species finds its southern limits in the Carolinas, and in our state, where it is listed as significantly rare, it is known only from boggy sites in two mountain counties. Root and leaf teas have diuretic, emetic, and expectorant effects, but the leaves are poisonous if eaten in large quantities. Marsh Marigold can be grown from seed for cooler climates; please do not disturb! ✗

APRIL–JUNE (76-6-1)
H-5B/SOD/SRY
[○/W/$/✪/△/SR]

Ranunculaceae

BANEBERRY; DOLL'S EYES

Actaea pachypoda Elliott

These native perennials are 4–8 dm tall, and each stem bears a compact raceme of small flowers. It is the fruit, however, that attracts attention—red pedicels bearing white berries with a dark dot (the doll's eye). The compound leaves have ovate, sharply toothed leaflets.

These plants are native in the rich woods of our mountains and piedmont and range throughout eastern North America. They can be grown from root sections, or from seed if you are not in a hurry. Native Americans used a baneberry root tea to relieve pain, but note that the roots, sap, and fruits are all highly poisonous!

APRIL–MAY (76-8-1)
H-5A/BOS/RRW
[●/M/$/✪/◉]

Ranunculaceae

BLACK COHOSH

Actaea racemosa Linnaeus
[*Cimicifuga racemosa* Nuttall]

The slender, graceful racemes of these native, rhizomatous perennials, up to 2.5 m tall, are 1–6 dm long; and, since the flowers open in succession upward, a raceme may remain in bloom for 2 weeks or more. The large, compound leaves are mostly basal.

Relatively frequent in clearings in rich woods on basic soils of our mountains and piedmont, these striking plants range through much of the eastern United States north of Florida. Black Cohosh can be grown from seed or rhizome cuttings and makes a good background for lower perennials in the wild garden. It has had wide use in Native American and folk medicine and is in active herbal use today for its estrogenic and anti-inflammatory activities.

MAY–JULY (76-9-1)
H-4B/BOS/RRW
[◗/M/$/✪]

Ranunculaceae

LEATHER FLOWER

Clematis viorna Linnaeus

This slender perennial vine has silky fruits that are often more conspicuous than the nodding, 1.5–2.5 cm long, light purple flowers, which are formed from the fused, thick, fleshy, petal-like sepals. The compound leaves have 3 to 7 leaflets, 2–8 cm long.

Chiefly of the mideastern to central United States, Leather Flower is found along the margins of rich woodlands on basic or limestone soils in our mountain and piedmont regions. Clematis is easily grown from seed in pots for the deck or patio, or in the wild garden. All Clematis species are poisonous if eaten.

MAY–SEPTEMBER (76-10-4)
V-4O/BOS/STR
[◗/M/$/✪/◉]

Ranunculaceae

VIRGIN'S BOWER

Clematis virginiana Linnaeus

This very showy native perennial vine has compound leaves with 3 leaflets. The very fragrant cream white flowers have separate, petaloid sepals, 6–10 mm long. As with other Clematis species, these plants are quite conspicuous in bloom as well as in fruit, when the wind-dispersed achenes are mature.

Virgin's Bower is found along streams and in openings in low woods in our piedmont and mountains and ranges throughout eastern and central North America. This species is a good candidate for the wild garden, where care should be taken to prevent unwanted spread. Although Virgin's Bower has been used medicinally, all parts of the plant are poisonous and will produce severe pain in the mouth if eaten.

JULY–SEPTEMBER (76-10-7)
V-40/POS/URW
[◗/M/$/⊙/◉]

Ranunculaceae

RUE ANEMONE

Thalictrum thalictroides (L.) Boivin

These low, delicate, early spring ephemeral perennials, 1–2 dm tall, have leaves that are twice to three times ternately compound. The flowers have 5 to 10 sepals, 5–18 mm long, that may be either pure white or tinged with pink.

A frequent inhabitant of rich, often low deciduous woods of most of the eastern and central United States, these plants are found chiefly in our mountains and piedmont. A tea made from the roots of Rue Anemone was used medicinally by Native Americans and has been used historically by physicians. Try a few of these plants in the rock garden.

MARCH–MAY (76-11-1)
H-VA/BOE/URW
[●/M/$/⊙]

Ranunculaceae

MEADOW RUE

Thalictrum revolutum DeCandolle

These erect, coarse, dioecious perennials, 0.5–1.5 m tall, are strikingly different from the small Rue Anemone. The many small flowers, about 5 mm long, are widely spaced and give the inflorescence a feathery appearance. The 0.5–3.5 cm wide leaflets are slightly revolute, or turned under, giving this plant its species name.

Found chiefly in the dry woods and meadows of our mountains and piedmont, this Meadow Rue ranges throughout much of the eastern and central United States. Seven other tall, somewhat similar species of Meadow Rue occur in North Carolina.

MAY–JULY (76-11-5)
H-VA/BOE/PRW
[◗/D/$]

Ranunculaceae

MOUSE-TAIL

Myosurus minimus L.

This small, inconspicuous annual, only 3–15 cm tall, is a true anomaly among wild flowers: the plants produce 2 to 20 scapes, or flower stalks, each 2–15 cm tall but each bearing only a single flower, with or without petals, near the base of an elongate receptacle (the "mouse tail"). The tail will grow up to 6 cm long in fruit.

Native in low, moist fields in scattered piedmont and coastal plain counties, Mouse-tail is circumboreal and, in America, primarily western, extending eastward into our area.

MARCH–MAY (76-12-1)
H-5B/SLE/SRG
[❍/M/✪]

Ranunculaceae

HOOKED BUTTERCUP

Ranunculus recurvatus Poiret

This low, hirsute perennial buttercup, 1.5–6 dm tall, is named for the strongly recurved beak or hook on each of the small, flat, clustered fruits. Both basal and stem leaves are mostly 3-lobed or -parted, and the flowers are approximately 10 mm in diameter.

These plants, widespread in the eastern and central United States, are found in rich, low woods throughout North Carolina except on the outer coastal plain. Not as showy as some of the other buttercups, it is still a good candidate for the wild garden. All parts of the plant are toxic if eaten in large quantities.

APRIL–JUNE (76-13-14)
H-5A/SOL/URY
[◗/M/✪/△]

Ranunculaceae

BUTTERCUP

Ranunculus bulbosus Linnaeus

These coarse, usually pubescent, erect perennials, 2–6 dm tall, from a cormose base, have palmately divided leaves, 3–10 cm wide. The numerous, waxy yellow flowers are up to 2.5 cm wide.

Found through much of the eastern United States, this introduced and now naturalized Buttercup grows in fields and meadows and along roadsides, where it can make quite a show. It occurs throughout our state but most commonly in the mountains and piedmont. The closely related *R. acris* differs only in its lack of corms. All Buttercups are easily grown from seed, but note that all parts of the plant are toxic when ingested.

APRIL–JUNE (76-13-17)
H-5A/POS/SRY
[○/N/✪/△]

Ranunculaceae

HEPATICA; LIVERWORT

Anemone acutiloba (DeCandolle) G. Lawson
[*Hepatica acutiloba* DeCandolle]

The new green, 3-lobed leaves that appear on these native perennials after the flowering period persist through the summer and following winter, by which time they have turned bronze or reddish-brown. The 4–15 cm tall flowering scape produces a solitary flower 6–10 mm long, with petaloid sepals that may be either pale blue, white, or rarely rose (as illustrated here).

Ranging throughout eastern and central North America south to Georgia and Missouri, Hepatica occurs rather frequently in the rich woods of western North Carolina. As the common name implies, a leaf tea made from either species of Hepatica found in North Carolina (*Anemone acutiloba* or *A. americana*) was used in both folk and Native American medicine for the treatment of liver ailments. Overcollection created an early threat to Hepatica populations in the nineteenth century, which may help account for the plant's relative scarcity today.

MARCH–APRIL (76-15-1)
H-VB/SOL/SRB
[●/M/$/✪]

Ranunculaceae

HEPATICA; LIVERWORT

Anemone americana (DeCandolle) H. Hara
[*Hepatica americana* (DC.) Ker]

Although similar in most respects to *A. acutiloba*, this plant has 3 leaf lobes that are rounded instead of pointed. The flowers are usually lavender but may be white or rose.

These native perennials occur over much of the eastern United States; they are relatively frequent in the rich woods of the North Carolina piedmont, but they are less frequent in our mountains and coastal plain.

FEBRUARY–APRIL (76-15-2)
H-VB/SOL/SRB
[●/M/$/✪]

Ranunculaceae

THIMBLEWEED

Anemone virginiana L.

These woodland perennials, with 3-parted basal leaves, have several white buttercuplike flowers on a stem 3–10 dm tall. The common name comes from the rounded, 1.5–3 cm long thimble-shaped fruiting heads made up of numerous, densely wooly, brownish achenes.

Thimbleweed occurs in rich woods through much of the eastern and central United States and, in our area, primarily in the mountains and piedmont and in calcareous areas of the coastal plain. Seeds and fruits of Thimbleweed were used as an expectorant, astringent, and emetic. Tolerant of partial shade, Thimbleweed makes an excellent choice for the native garden but note that fresh sap will cause blistering and, if eaten, severe pain in the mouth.

MAY–JULY (76-16-1)
H-50/SRL/URW
[◗/M/$/✪/⚠]

Ranunculaceae

WOOD ANEMONE

Anemone quinquefolia Linnaeus

The flowers of these 0.5–3.5 dm tall, rhizomatous, native perennials have firm white sepals 8–16 mm long. The rounded cluster of small, densely wooly fruits is about 1 cm in diameter. The stems and leaves are usually glabrous.

Primarily of northeastern North America, Wood Anemone ranges south through the mountains, where, in our area, it is found in rich woods and along woodland borders at scattered localities. Plants are toxic when eaten in large quantities.

MARCH–MAY (76-16-3)
H-VB/SOC/SRW
[●/M/$/⚠]

Berberidaceae

TWINLEAF

Jeffersonia diphylla (L.) Persoon

The strongly 2-parted leaf on a 1.5–3 dm petiole, which looks like a green butterfly with its wings spread, offers immediate and positive identification of this rare woodland perennial—and gives it both its common and scientific names. The solitary flower on the 1–2 dm tall scape has eight 1–2 cm long petals. The fruit is a follicle that splits when ripe to release numerous seeds that are dispersed by ants.

These plants grow in rich, deciduous woods on basic soils in only one mountain county, but the species ranges more broadly in the eastern and central United States. While listed in North Carolina, there is no federal listing for this species since it is not rare throughout its range. The only other species in the genus (named for Thomas Jefferson) occurs in eastern Asia. The Cherokee and Iroquois used whole plant infusions and root tea of Twinleaf for medicine. ✗

MARCH–APRIL (77-4-1)
H-8B/PRE/SRW
[●/M/$/✪/SR]

Berberidaceae

BLUE COHOSH

Caulophyllum thalictroides (L.) Michaux

These glabrous native perennials, 3–8 dm tall, have a single, large, divided leaf and a single flower stalk. The bluish stem and leaves account for the common name. The small greenish-yellow flowers are less conspicuous than the small, blue, exposed seeds, which grow faster than the ovary and split it open.

Broadly distributed in northeastern and central North America, Blue Cohosh ranges south to Georgia and west to Oklahoma and is found in our area in the rich deciduous forests of mountain coves and at two localities in the piedmont near the center of the state. Blue Cohosh has a long historical record of medicinal use and is in common use today as a homeopathic medicine. Both the seeds and roots are poisonous.

APRIL–MAY (77-5-1)
H-6B/BBS/PRG
[●/M/$/✪/△]

Berberidaceae

UMBRELLA-LEAF

Diphylleia cymosa Michaux

This rhizomatous perennial produces a single, large, umbrellalike leaf, 2–5 dm across, on nonflowering stems and two somewhat smaller leaves on flowering stems. The white flowers, each about 15 mm broad, are followed by round blue fruits, 1 cm in diameter, on stalks 4–10 dm tall.

These rare but conspicuous plants are found on rich seepage slopes only in the southern Appalachians and are infrequent to rare in our westernmost mountain counties. Propagation by seed can produce flowering plants in 4 to 5 years—if your climate is cool enough. ✗

MAY–JUNE (77-6-1)
H-6B/SRL/URW
[●/W/$/✪/R]

Berberidaceae

MAYAPPLE; MANDRAKE

Podophyllum peltatum Linnaeus

The flowering stems of these 3–5 dm tall rhizomatous perennials have 2 roughly circular, 3–7 lobed, peltate leaves, 3–4 dm wide, with a single, waxy, white flower, 3–5 cm across, between them. Flowerless stems have only a single leaf.

Native to the eastern and central United States, Mandrake is frequent in low, alluvial woods and moist meadows throughout North Carolina. They are easily propagated from rhizome sections and form large colonies through rhizomatous growth. But note that clones are self-sterile, so be careful to collect rhizomes from more than one colony if you want fruit production. The plants have both historical and modern use as a medicinal, but the root, leaf, and unripe fruit are all highly toxic.

MARCH–APRIL (77-7-1)
H-6B/SRL/SRW
[◗/M/$/✪/◉]

Menispermaceae

CORALBEADS

Cocculus carolinus (L.) DeCandolle

This slender, dioecious, perennial vine with small, greenish white flowers has alternate, weakly or strongly lobed leaves 2.5–12 cm long. The translucent, crimson berries, 5–8.5 mm in diameter, ripen in the early fall.

These vines, native to the southeastern United States, are found in our area along fencerows and in fields and sandy woodlands of the coastal plain and piedmont. Coralbeads is a good choice to plant along a fence.

JUNE–AUGUST (79-1-1)
V-VA/SOL/PRG
[◗/N/$/☉]

Magnoliaceae

TULIP TREE

Liriodendron tulipifera Linnaeus

The tulip-shaped flowers with brilliant orange and green petals, 3.5–6 cm long, and the wide leaves, 6–20 cm long and wide with a shallow notch at the end, quickly identify this large, handsome, fast-growing tree.

Native to the eastern United States, Tulip Trees are found in coves or low, often recently cleared woodlands, essentially throughout the state. The wind-dispersed seeds have enabled these trees, in many areas, to replace the Chestnut trees killed by the blight. Tulip Trees are commercially valuable for their timber and also make great landscape trees, which, with a century or so to grow, can reach a diameter of over 3.5 m and a height of up to 60 m. Native Americans used the bark, leaves, and buds for medicine.

APRIL–JUNE (80-1-1)
T-6A/SRL/SCY
[◗/M/$/☉]

Magnoliaceae

SWEET BAY

Magnolia virginiana Linnaeus

This is usually a medium-sized to small tree, up to 20 m, or, in areas that have been frequently cut over or burned, a bushy stump sprout. The semi-evergreen leaves, whitish beneath, are 6–15 cm long, and the fragrant, cup-shaped white flowers are approximately 5 cm across.

A native of the eastern United States from New York to Florida and along the Gulf Coast west to Texas, Sweet Bay is frequent to common on savannas, in pocosins, and in roadside thickets, chiefly of our coastal plain. Sweet Bay is an easy-to-grow small landscape tree. A tea made from its bark was used to treat fever, rheumatism, and indigestion.

APRIL–JUNE (80-2-1)
T-VA/SEE/SCW
[◗/M/$/✪]

Magnoliaceae

MAGNOLIA; BULL BAY

Magnolia grandiflora Linnaeus

This large, handsome evergreen tree, up to 30 m tall, has leaves 1–3 dm long that are glossy above and reddish-brown pubescent beneath. The fragrant flowers, pollinated by beetles, may be 3 dm across when fully open.

Although widely planted over a much larger area, this native of the southeastern United States reaches its natural northern limit in the swamp forests and low woods of southeastern North Carolina.

MAY–JUNE (80-2-2)
T-VA/SEE/SCW
[◗/M/$/✪]

Magnoliaceae

UMBRELLA TREE; MOUNTAIN MAGNOLIA

Magnolia fraseri Walter

The thin, deciduous leaves of this slender native tree, up to 15 m tall, are 1–5 dm long and have 2 prominent lobes (or "ears") at the base. The petals are 8–12 cm long. The leaves of the somewhat similar *Magnolia tripetala*, found scattered over the state, are not notched at the base.

A native primarily of the southern Appalachians, this Magnolia is found in rich woods of the mountains and upper piedmont of North Carolina and, in its natural range, makes a great landscape tree, as does the easily grown *M. tripetala*.

APRIL–MAY (80-2-4)
T-VA/SBE/SCW
[◗/M/$]

Annonaceae

PAWPAW

Asimina triloba (L.) Dunal

This slender, often shrubby native tree, up to 10 m tall, has thin, obovate leaves that are odorous when bruised. The leaves, 15–25 cm long, appear after the maroon 3-parted, 3–4 cm broad flowers. The fleshy, fragrant fruit, 6–15 cm long, is edible, but woodland animals usually get it first!

Native to the eastern and central United States, these trees are infrequent in low woods at scattered localities throughout the state. Easily grown from seed, these plants, with good fall color, make interesting landscape subjects. Warning: the seeds and leaves are toxic if ingested.

MARCH–MAY (81-1-2)
T-6A/SBE/SCM
[◗/M/$/✪/△]

Calycanthaceae

SWEET SHRUB; SWEET BETSY

Calycanthus floridus Linnaeus

This 1–3 m tall native shrub, has opposite, deciduous leaves, 5–18 cm long, that turn bright yellow in the fall. The brown-maroon flowers are 3–4 cm broad and have a spicy fragrance; the "seeds," found in fall in a papery urn-shaped receptacle, are actually fruits and are usually easy to germinate.

A plant of the southeastern United States, this shrub is found along roadsides and stream banks on the margins of deciduous woodlands, chiefly in our upper piedmont and mountains. Easily propagated by seed or cuttings, these attractive shrubs are planted as an ornamental and are highly recommended for landscape use. Warning: the "seeds" are poisonous.

MARCH–JUNE (83-1-1)
S-VO/SNE/SRM
[◗/N/$/✪/△]

Lauraceae

SASSAFRAS

Sassafras albidum (Nutt.) Nees von Esenbeck

The paniculate clusters of small yellow flowers appear before the leaves on these small trees or shrubs, usually about 10–15 m tall. However, Sassafras is most easily recognized by its distinctive, 6–12 cm long leaves that have either 3 lobes, 2 lobes or no lobes, all on the same tree. The striking, fleshy blue-black fruits with red stalks ripen in midsummer but soon disappear, as they are extremely popular forage for fruit-eating birds. In the fall, Sassafras leaves provide striking color, usually a brilliant bronze-orange but varying from bright red to clear yellow.

This native of the eastern and central United States is found throughout our area, frequently as a low tree in cutover areas along fencerows and in open woodlands. The aromatic root bark tea has long been used as a spring tonic or "blood purifier." More recently, it has been used to flavor root-beer, but there is some indication that the active ingredient may be carcinogenic, so it is no longer used. The finely powdered, dried young leaves, or "filé," contribute to the mucilaginous gumbo of Louisiana cooking. The aromatic, insect-repellent wood was once used for bedposts and chicken roosts.

MARCH–APRIL (84-2-1)
T-6A/SEL/PRY
[◗/N/$/✪/△]

Lauraceae

SPICEBUSH

Lindera benzoin (L.) Blume

These deciduous, aromatic, 1–3 m tall shrubs have axillary clusters of small yellow flowers along the branches in early spring before the leaves come out. The thin, obovate leaves, 6–14 cm long, turn a brilliant yellow in the fall and contrast sharply with the bright red berries.

Spicebush ranges throughout the eastern United States and is found in basic soils of alluvial woodlands in all three provinces of North Carolina. The bright fall foliage and the crimson fall berries make this a great shrub for the garden and for landscape use. Wood Thrushes and Veerys, and few other birds, eat the fruits. Native Americans made extensive use of the leaves, fruits, twigs, and bark for medicine. This is a shrub that, as a landscape or wild garden plant, "has everything"!

MARCH–APRIL (84-4-1)
S-6A/SBE/URY
[◗/M/$/⊙]

Papaveraceae

BLOODROOT

Sanguinaria canadensis Linnaeus

The rhizome of this low native perennial, 1–4 dm tall, has a bright, red-orange sap that gives the plant its name. The solitary, showy flowers of these spring ephemerals have 8 to 12 petals, 1.5–3 cm long. The usually solitary leaf, glaucous beneath, is 5–18 cm wide and deeply cleft into numerous wide segments. Ants, attracted by the fleshy white aril, disperse the seed across the forest floor, harvesting the aril and leaving the seed to germinate.

A native of eastern and central North America, Bloodroot is frequently found in open mixed deciduous forests, chiefly of our mountains and piedmont. Bloodroot is readily propagated by seed or rhizome sections for the shade garden. Used historically in medicine and as a decorative skin paint, the bright red, bitter alkaloid sanguinarine found in the rhizome also has modern uses, but note that these rhizomes are highly poisonous. Do not ingest!

MARCH–APRIL (85-1-1)
H-8B/SRL/SRW
[●/N/$/✪/◉]

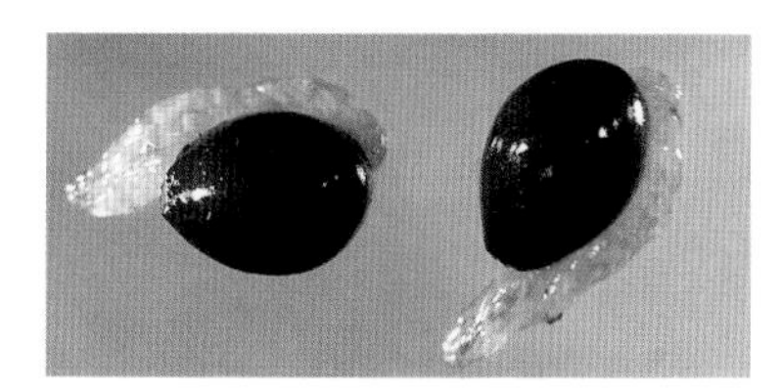

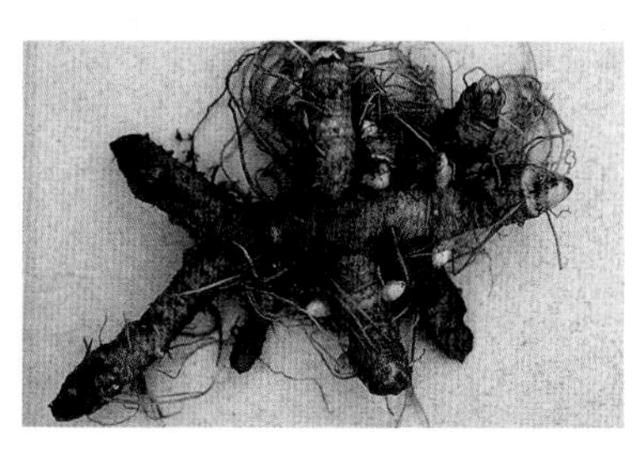

Papaveraceae

POPPY

Papaver dubium Linnaeus

The showy red, rarely white, 3–7 cm wide flowers of this 3–6 dm tall, introduced annual last hardly a day. The pinnately cut leaves are 4–10 cm long and the smooth, club-shaped capsule is about 2.5 cm long.

Though infrequent in fields and on roadsides at a dozen scattered localities in our state, these colorful plants appear to be thoroughly naturalized in the vicinity of Wilmington where, the story goes, they were once grown commercially.

APRIL–JUNE (85-5-4)
H-4A/SLC/SRR
[○/N/$]

Fumariaceae

BLEEDING HEART

Dicentra eximia (Ker) Torrey

A low perennial, 2–5 dm tall, with drooping, flattened pink or rose flowers 2–2.5 cm long. The ovate leaves, 5–25 cm long and wide, are twice pinnately compound.

These rare plants, with a state protected status, grow on rich, wooded slopes and in the coves and gorges of only a few of our mountain counties and have a general range centered in the Appalachians. Bleeding Heart is often grown in gardens and is widely cultivated. The whole plant may be a skin irritant and is poisonous if ingested. ✗

APRIL–JUNE (86-2-1)
H-4B/BOL/PTR
[◗/N/$/⚠☞/*SR*]

Fumariaceae

SQUIRREL CORN

Dicentra canadensis (Goldie) Walpers

The roots of this low native perennial, 1–3 dm tall, form small, round yellow tubers—thus the common name. Although the finely divided leaves of Squirrel Corn are similar to those of Dutchman's Breeches, the rounded, approximate basal spurs of the 12–18 mm flowers serve to distinguish this species.

Primarily a northern plant of the eastern and central states south in the mountains to North Carolina, it is rare on moist slopes and in wooded coves, occasionally in large stands. Like other *Dicentra* species, Squirrel Corn is poisonous. ✗

APRIL–MAY (86-2-2)
H-4B/BOL/RTW
[●/M/$/⚠☞/*R*]

Fumariaceae

DUTCHMAN'S BREECHES

Dicentra cucullaria (L.) Bernhardi

Although vegetatively similar to Squirrel Corn, the basal spurs of the flowers of this species are half or more the total flower length and are also more divergent. The flowering scapes are 1–3 dm tall.

Primarily a northern plant of the eastern and central states south in the mountains to Georgia, Dutchman's Breeches is found infrequently in rich woods and on north slope river banks in our mountains and piedmont. Of the three *Dicentra* species occurring in North Carolina, this is the only one with records of medicinal use; the whole plant is poisonous if ingested.

MARCH–APRIL (86-2-3)
H-4B/BOL/RTW
[●/M/$/✪/⚠☞]

Fumariaceae

PALE CORYDALIS

Corydalis sempervirens (L.) Persoon

Named for the glaucous or pale character of the finely divided foliage, these native, biennial plants are 3–7 dm tall. The flowers are 10–15 m long.

Ranging across Canada and the northern United States south through the mountains to Georgia, this species is infrequent to rare on open rocky slopes and outcrops in our mountains and a few piedmont counties. Pale Corydalis is frequently cultivated in the wild garden.

APRIL–MAY (86-3-1)
H-4A/POL/RZR
[○/D/$/✪]

Brassicaceae

POOR MAN'S PEPPER; PEPPERGRASS

Lepidium virginicum L.

This erect winter annual, from a basal rosette, is 1–5 dm tall; the rosette leaves, often highly dissected and serrate, are not present at flowering. The stem is branched from the upper axils, and the tiny white flowers are succeeded by far more conspicuous flat, round, pungent fruits that are 3–3.5 mm long.

This common weed of fields, gardens, and other disturbed habitats is found throughout the United States and southern Canada. The fruits of Peppergrass have been used to season salads, and Native Americans made use of both its fruits and seeds for medicine.

APRIL–JUNE (88-4-3)
H-4A/SED/RRW
[○/N/✪]

Brassicaceae

SEA ROCKET

Cakile edentula (Bigelow) Hooker

These low, succulent, branched annuals, 1–5 dm tall, have thick, alternate leaves 3–10 cm long and 1–4 cm wide. Plants flower sporadically throughout the year; each flower is 10–15 mm in diameter. The distinctive common name comes from the fruits, each with two segments (containing a single seed), that resemble a two-stage rocket and are adapted for water dispersal.

Sea Rocket grows on dunes and beaches in our outer coastal plain and more broadly along both the eastern and western coastlines of the United States and the Great Lakes states.

MAY–JUNE (88-8-2)
H-4A/SED/RRW
[○/D]

Brassicaceae

DAME'S ROCKET

Hesperis matronalis L.

This naturalized biennial or weak perennial has stems 1–1.5 m tall, with showy racemes up to 5 dm long of lavender or purple, rarely white, flowers 1.5–2 cm long.

Dame's Rocket occurs in only a few of our mountain counties but is a common introduction or escape in much of the central and northern United States. Very easily grown from seed and tolerant of partial shade, these plants are widely available for the garden. Note, however, that Dame's Rocket can be moderately invasive.

MAY (88-17-1)
H-4A/SND/RRB
[◗/N/$]

Brassicaceae

WINTER CRESS

Barbarea verna (Miller) Ascherson

The branched stems, 3–8 dm tall, have pinnately dissected leaves up to 15 cm long and yellow 4-petaled flowers 1–1.5 cm in diameter. Several named varieties of these naturalized, widespread winter annuals, or biennials, have been described on the basis of variations in fruit length and the number of lobes on the leaves, but they all look, and taste, essentially the same.

Winter Cress is common in disturbed, moist habitats throughout our state and generally throughout eastern North America. The leaves or whole plant of Winter Cress can be used raw as a salad or cooked as "greens."

MARCH–JUNE (88-22-1)
H-4A/SLC/RRY
[○/N/$]

Brassicaceae

CUTLEAF TOOTHWORT

Cardamine concatenata (Michx.) O. Schwarz
[*Dentaria laciniata* Willdenow]

This native, 2–4 dm tall perennial from a small, deeply set, fleshy tuber, or jointed rhizome, usually has 3 stem leaves, deeply parted or divided into 3 to 5 segments 0.5–3 cm wide. The cruciform, 12–19 mm long flowers are followed by the narrow, round capsules, or siliques, which are 1–4 cm long.

A species of eastern North America, Cutleaf Toothwort is relatively frequent in alluvial woods at scattered localities across North Carolina, though it is most abundant in the mountains and piedmont. Three other Toothworts occur in the state, and all are good candidates for the wild garden. The Iroquois used the roots for both food and medicine.

MARCH–MAY (88-23-3)
H-4W/PLL/RCW
[●/M/$/✪]

Brassicaceae

BITTER CRESS

Cardamine clematitis Shuttleworth

A rhizomatous perennial with erect, pubescent stems, 2–4 dm tall, this native Bitter Cress has pinnate leaves with obovate, dentate leaflets. The 5–10 mm long, cruciform flowers are in terminal racemes.

Native only to the southern Appalachians, Bitter Cress is found in North Carolina growing by, or in, rocky streams in the mountains. Plants of some species, as the common name implies, are used for cress, or salad.

APRIL–MAY (88-23-7)
H-4A/POD/RCW
[●/M/$]

Sarraceniaceae

YELLOW PITCHER PLANT; TRUMPETS

Sarracenia flava Linnaeus

Insects are trapped in the long, slender, hollow leaves of these interesting perennials, where they are digested and provide the plants with some of their mineral requirements; the leaves may be 3–10 dm tall. The flowers, with 5 obovate, 5–8 cm long petals and an interesting, flattened, style disc, have a very strong musky scent and generally appear with or before the new leaves.

A native of the southeastern United States that is becoming quite rare on savannas and in bogs and wet ditches of the coastal plain. Do not dig! Grow these plants from seed; you will have tiny 2–3 cm tall pitchers the first year. ✗

MARCH–APRIL (89-1-1)
H-5B/SLE/SRY
[○/W/$/R]

Sarraceniaceae

SWEET PITCHER PLANT

Sarracenia rubra Walter

The slender flowering stems of these plants are nearly twice as long as the narrow, 1–5 dm tall, hollow leaves. The ovate opening of the pitcher is 1–3 cm broad. The maroon flowers, with strongly obovate petals, 2–3 cm long, have the fragrance of violets and are smaller than those of North Carolina's other *Sarracenia* species.

A rare, rhizomatous perennial of the southeastern United States still found in a few shrub bogs and savannas of the coastal plain, Sweet Pitcher Plant appears to be extinct at several previously known localities along mountain streams. The plant reaches its northern limit in our state. Buy nursery-grown plants or grow your own; do not dig! Watch the seedlings grow into mature Pitcher Plants for your bog garden. ✗

APRIL–MAY (89-1-2)
H-5B/SLE/SRM
[❍/W/$/*R*]

Sarraceniaceae

HOODED PITCHER PLANT

Sarracenia minor Walter

The spotted, hooded, hollow leaves of these plants are 2–4 dm tall, about the same height as the flowering stalk. The leaves gradually expand toward the opening, over which the hood arches closely. The flowers are odorless; the petals are 2.5–4 cm long.

This is another southeastern species that reaches its northern limit in North Carolina, where it is state protected. These rhizomatous perennials are now rarely seen in savannas of only two counties of our southeastern coastal plain. This Pitcher Plant is wonderful to grow from seed for the bog garden; the beauty is worth the three-year wait. Note: there are a number of fantastic hybrids now available in the trade. ✗

APRIL–MAY (89-1-3)
H-5B/SBE/SRY
[❍/W/$/*SR*]

Sarraceniaceae

PITCHER PLANT; FLYTRAP

Sarracenia purpurea Linnaeus

The open, hollow leaves of this Pitcher Plant are 0.5–3 dm or more long, widest near the middle, and form a basal rosette around the flower stalk. The hood is erect so that the opening is completely exposed and the pitchers are often filled with water and decaying insects. The flowering stalk is 2–4 dm tall; the odorless flowers have maroon petals 4–6 cm long.

A now rare but once widespread species of the eastern United States and Canada, these rhizomatous perennials are still found in a few bogs and moist savannas of the coastal plain and at a few similar isolated localities in the mountains. This is the only Pitcher Plant occurring in North Carolina that has been used extensively for medicine to treat smallpox and lung and liver ailments and as a childbirth aid. Prepackaged or mail-order plants never survive; grow your own from seed or insist on nursery-grown plants. The plants are long-lived and, in a stable, acid, sphagnum bog, can become quite large. ✗

APRIL–MAY (89-1-4)
H-5B/SEE/SRM
[◐/W/✪/R]

Crassulaceae

SEDUM; STONECROP

Sedum ternatum Michaux

This prostrate, mat forming native perennial usually has a single flowering stalk, 5–15 cm tall, and several sterile shoots. Basal cauline leaves on the sterile and flowering stalks are in whorls of 3. The small, rounded, fleshy leaves are 0.7–3 cm long.

Occurring through much of the eastern United States, this Sedum is found scattered in rich, rocky, but often moist woods of our piedmont and mountains. This plant makes a great groundcover for the garden.

APRIL–JUNE (91-1-5)
H-5A/SOE/URW
[◐/M/$/△]

Crassulaceae

SEDUM; LIVE-FOR-EVER

Sedum telephioides Michaux

This is North Carolina's most distinctive species of Sedum, with wide, fleshy, petiolate leaves 3–6 cm long on an erect stem 2–5 dm tall.

These succulent perennials, generally of the Mid-Atlantic states, grow in rock crevices and rocky woods of the North Carolina mountains and a few adjacent piedmont counties. They are good plants for the rock garden.

JULY–SEPTEMBER (91-1-9)
H-5A/SBS/URW
[❍/D/$/✪/△]

Droseraceae

THREADLEAF SUNDEW

Drosera filiformis Raf.

The 14–25 cm long, filiform leaves of these cormose perennials bear many glandular hairs and are circinate (coiled in bud)—the leaf actually "unrolls" as it grows. Flowers are borne in racemes 4–15 cm long on scapes about as long as the leaves. Flies and other small insects are caught by the shining, sticky droplets of mucilaginous material produced at the tips of the slender red hairs on the leaves and provide the plant with important nutrients to supplement nutrient-poor soils.

These interesting but quite rare, state-listed plants grow in wet savannas and bog margins in only three of our coastal counties and at scattered areas along the coast, essentially from Massachusetts to Mississippi. ✗

JUNE (92-1-1)
H-5B/SLE/RRR
[❍/M/$/*SR*]

Droseraceae

SUNDEW

Drosera rotundifolia Linnaeus

The rounded leaf blades of these small perennials are about as wide as long, up to 9 mm, thus the specific epithet. The glistening sticky droplet on each leaf hair accounts for the common name, Sundew. The flowering scape is 10–18 cm long; the raceme, 1–6 cm long.

These plants grow in the sphagnum bogs of a few of our mountain counties. Though rare in North Carolina, they are more common northward, where they are circumboreal, extending south in the mountains to Georgia. The whole plant has been used traditionally in teas and tinctures for lung ailments. ✗

JULY–SEPTEMBER (92-1-3)
H-5B/SRE/RRW
[❍/w/$/✪/R]

Droseraceae

SPOONLEAF SUNDEW

Drosera intermedia Hayne

The stem or scape of this Sundew is 5–10 cm tall; the leaves are in a basal rosette and also at intervals on the flowering stem. The cauline leaves are spoon-shaped, 7–15 mm long, with long petioles, three to four times as long as the blade. The minute flowers stay open only a few hours.

Ranging throughout eastern North America, this interesting perennial is relatively frequent in our coastal plain, where it grows in standing water in bogs and wet ditches. ✗

JULY–SEPTEMBER (92-1-4)
H-5A/SBE/RRW
[❍/w/$]

Dionaeaceae

VENUS FLYTRAP

Dionaea muscipula Ellis

This is one of the world's most unusual plants. The red color inside the 2-lobed leaf of these small native perennials attracts flies and other insects that are then trapped when the lobes, up to 2.5 cm long, suddenly close together. The white flowers are in an umbellate inflorescence at the top of a flowering scape 1–3 dm tall.

A very rare plant, originally known only from the coastal area of the Carolinas, Venus Flytrap is now extinct (or extirpated) at many localities where it previously grew in open sun in wet savannas, seepage bogs, and pocosin edges. It is both state and federally listed. Do not dig; grow from seed! ✗

MAY–JUNE (93-1-1)
H-5B/SOL/URW
[○/W/$/*SR-SC*]

Saxifragaceae

VIRGINIA WILLOW

Itea virginica L.

This attractive, 1–2 m tall shrub has alternate, elliptic or obovate leaves 2–9 cm long. The terminal racemes of numerous small, cup-shaped flowers are 4–15 cm long.

Widely distributed in North Carolina along streams and swamp margins and in low woods, Virginia Willow ranges generally throughout the Southeast to Texas. This shrub is widely available and makes a great addition to a wild flower garden, where it tolerates shade but does best with several hours of full sun.

MAY–JUNE (94-1-1)
S-5A/SES/RCW
[◗/M/$]

Saxifragaceae

MOCK ORANGE

Philadelphus hirsutus Nuttall

There are over 50 varieties of this showy native shrub in cultivation, many with very fragrant flowers that are up to 4 cm across. The shrubs are 1–2 m tall, and the leaves, which are pubescent beneath, are 4–8 cm long.

A rare plant of the southern Appalachian region, this shrub is occasionally found on dry wooded bluffs and ledges in some of our mountain counties. ✗

APRIL–MAY (94-4-2)
S-40/SOS/URW
[◯/D/$/R]

Saxifragaceae

WILD HYDRANGEA

Hydrangea arborescens Linnaeus

This native, spreading shrub, 1–3 m tall, has numerous cultivated varieties. The white, showy, 3- or 4-lobed calyx of the outer, sterile flowers is 1–3.5 cm broad; the compact center of the inflorescence is made up of small fertile flowers. The ovate, opposite leaves are up to 18 cm long and 12 cm wide; subspecies are separated based on the leaf undersurface, which may be white, gray, or green.

Native to the eastern and central United States north to New York, these shrubs grow on shady, often moist road banks and cliffs of our mountains and upper piedmont and are easily propagated by cuttings. Native Americans used a Wild Hydrangea root tea as a diuretic, cathartic, and emetic, and poulticed bark on wounds and burns. Note: the bark, leaves, and flower buds are poisonous if ingested in large amounts.

MAY–JULY (94-5-1)
S-50/SOS/URW
[◗/M/$/✪/△]

Saxifragaceae

GRASS-OF-PARNASSUS

Parnassia asarifolia Ventenat

These native perennials, 1–5 dm tall, are easily identified by the prominent green veins on the 5 white, 10–16 mm long petals. The kidney-shaped basal leaves are 2.5–6 cm long.

This species is native to the southern Appalachians and is rather rare in North Carolina, where it grows in bogs and on seepage slopes in the mountains. Two similar species occur in the state, *P. grandifolia* in the mountains and *P. caroliniana* in the coastal plain; both are endangered and are state listed. Grass-of-Parnassus can be grown from seed and is highly recommended for open, wet areas of the wild garden. Do not dig it from the wild! ✗

AUGUST–OCTOBER (94-6-1)
H-5B/SCE/SRW
[◐/W/$/R]

Saxifragaceae

FALSE GOAT'S BEARD

Astilbe biternata (Vent.) Britton

These coarse, rhizomatous perennials, up to 1.5 m tall, have alternate, 2–3 ternately compound, basal and cauline leaves, 8–15 cm long. The terminal leaflets are generally 3-parted, unlike those of the dioecious true Goat's Beard (*Aruncus dioicus* in the Rose family), which is often confused with this species.

False Goat's Beard is infrequent in rich woods and on seepage slopes in our mountains and in the surrounding Mid-Atlantic states. Astilbe makes a great addition to the shade garden.

MAY–JULY (94-7-1)
H-5A/BOS/PRW
[●/M/$]

Saxifragaceae

MITERWORT; BISHOP'S CAP

Mitella diphylla L.

These rhizomatous perennials, 2–6 dm tall, have lobed, mostly basal leaves, 4.5–8 cm long. The slender flower stalk bears 5 to 20 perfect, intricate white flowers with 5 fimbriate petals, 2–2.5 mm long, which do indeed look like bishops' caps. These easily overlooked flowers deserve a second look—with a hand lens!

Mitella is found in rich woods of our mountains and generally through the Appalachians north to Canada. These plants, along with Foamflower and Alumroot, make wonderful additions to the wild flower garden.

APRIL–JUNE (94-10-1)
H-5B/SOL/RCW
[●/M/$/✪]

Saxifragaceae

HAIRY ALUMROOT

Heuchera villosa Michaux

The ovate, sharply lobed basal leaves of this native, rhizomatous perennial are 3–18 cm long and wide. The flowering stems are 2–7.5 dm tall. The tiny, tubular flowers, 3–4 mm long, have exserted stamens (protruding beyond the tube) and are arranged in a branched panicle that is usually rusty-villous, with long, shaggy hairs.

Hairy Alumroot grows on shaded rocks and ledges in our mountains and more broadly through the mountains from New York to Georgia. This plant makes a great addition to the rock garden.

JUNE–SEPTEMBER (94-11-1)
H-5B/SRL/PTR
[◗/N/$]

Saxifragaceae

MITERWORT; FOAMFLOWER

Tiarella cordifolia Linnaeus

This rhizomatous perennial has roughly heart-shaped, semi-evergreen leaves up to 14 cm long. The small, white, 5-petaled flowers are in showy racemes, 3–15 cm long, on flowering stalks 2–5 dm long. Two varieties are found in North Carolina, var. *cordifolia*, with stolons, and var. *collina*, without stolons.

These low herbs, native generally to the northeastern United States, are relatively frequent in the rich woods of our mountains and piedmont. Older plants of var. *collina* may form large clumps or colonies with many flower stalks that can be easily divided. Plants may also be grown from seed, making this species an excellent choice for the wild garden. Native American and traditional use of this plant to treat mouth sores, eye ailments, and wounds may be explained by its high tannin content.

APRIL–JUNE (94-12-1)
H-5B/SOL/RCW
[●/M/$/✪]

Saxifragaceae

MICHAUX'S SAXIFRAGE

Saxifraga michauxii Britton

These native perennials, 1–5 dm tall, have a basal rosette of often colorful, green to red, coarsely serrate, obovate, or oblanceolate leaves up to 18 cm long and 4 cm wide. The widely branching flower stalk bears small, white, zygomorphic flowers.

Limited generally to the southern Appalachians, these plants grow in the crevices of moist rocks and on seepage slopes in our mountains.

JUNE–AUGUST (94-14-1)
H-5B/SBS/PZW
[◗/W/$]

Saxifragaceae

MOUNTAIN LETTUCE

Saxifraga micranthidifolia (Haw.) Steudel

This native perennial Saxifrage, 3–8 dm tall, has lanceolate to oblanceolate leaves up to 3.5 dm long and 6 cm wide. Unlike those of *S. michauxii*, the small flowers in its paniculate inflorescences are regular.

Found in the cooler Appalachian Mountains from Pennsylvania to North Carolina, Mountain Lettuce grows on moist rocks and seepage slopes. As the common name indicates, the leaves of this Saxifrage have been used as a green by the Cherokee.

MAY–JUNE (94-14-4)
H-5B/SBS/PRW
[◗/W/$]

Hamamelidaceae

WITCH HAZEL

Hamamelis virginiana Linnaeus

This deciduous native shrub or small tree, up to 5 m tall, has toothed or shallowly lobed, oblique-based leaves, 5–15 cm long, which are more or less round. The small yellow flowers with 4 linear petals, 1.5–2 cm long, are in compact clusters just above the old leaf scars and appear on naked branches in the fall or winter. The brown capsules, 1–1.5 cm long, release their seeds explosively.

Witch Hazel is found in rich woods and along dry woodland margins throughout the state and eastern North America. An extract of bark, twigs, and leaves has had a long history of medicinal use, and Witch Hazel is still in commercial production today.

OCTOBER–DECEMBER (95-2-1)
S-4A/SOS/URY
[◗/N/$/✪]

Hamamelidaceae

WITCH ALDER; FOTHERGILLA

Fothergilla major (Sims) Loddiges

This native monoecious, colonial shrub, 0.5–1.5 m tall, has compact, attractive, 1.5–6 cm long spikes of small, apetalous flowers with numerous white filaments. The leaves, up to 12 cm long and 5 cm wide, turn brilliant yellow or red in the fall.

This narrow endemic of dry woods and balds is found in the piedmont and mountains of the Carolinas, Tennessee, Alabama, and Georgia, and is listed as threatened in our state. A lower-growing species, *F. gardenii*, is found in our coastal plain. Fothergilla makes an excellent landscape shrub. ✗

APRIL–MAY (95-3-1)
S-NA/SOS/IRW
[◐/D/$/*SR-T*]

Rosaceae

WILD STRAWBERRY

Fragaria virginiana Duchesne

This low native perennial often forms large colonies by means of stolons or runners. The leaf is made up of 3 obovate, dentate leaflets, each usually 2.5–12 cm long. The sweet, fleshy red fruits are often more flavorful than the related, but larger, cultivated strawberries.

A frequent to common plant of old fields and woodland borders throughout the state and much of North America, it can be propagated from the rooted runners. Native Americans utilized the leaves and roots for medicine.

MARCH–JUNE (97-1-1)
H-5B/POS/URW
[◐/N/$/✪]

Rosaceae

MOCK STRAWBERRY; INDIAN STRAWBERRY

Duchesnea indica (Andr.) Focke

Similar to the true strawberry, this low (3–10 cm tall), introduced, stoloniferous perennial has trifoliate leaves, but the petals are yellow and the pulpy fruit tasteless. Unlike the true strawberry, the fruit is surrounded by 5 large, toothed bracts.

An introduced Asiatic weed, Mock Strawberry is found in lawns, pastures, and open woods and on roadsides at scattered localities throughout our state and much of the eastern United States. Mock Strawberry has a rich medicinal history in Asia, where it was used as a treatment for lung ailments, as an astringent, and to promote increased blood circulation.

FEBRUARY–FROST (97-2-1)
H-5B/POS/URY
[◯/N/☉]

Rosaceae

CINQUEFOIL; FIVE FINGERS

Potentilla canadensis Linnaeus

This stoloniferous native perennial has 5 unequal leaflets, the middle leaflet up to 4 cm long. The young leaflets, petioles, and lower leaf surface are pubescent, with long, silky hairs. The solitary flowers are on axillary stalks, 1.5–6 cm long, elongating in fruit to 5 dm.

A northeastern species extending southward to Georgia, these often weedy plants are relatively frequent in pastures, lawns, and dry woodlands more or less throughout our state. Many *Potentilla* species have been used medicinally for their astringent properties.

MARCH–MAY (97-4-1)
H-5A/PRS/SRY
[◗/N/☉]

Rosaceae

SULPHUR CINQUEFOIL

Potentilla recta L.

These 4–8 dm tall introduced perennials have pubescent, palmately compound leaves with 5 to 7 oblanceolate, coarsely serrate leaflets, 2–10 cm long. The light yellow flowers, 1–1.5 cm in diameter, are in compound cymes.

This Eurasian weed is found in relatively dry soil in pastures and along roadsides and railroads throughout the state and has spread sporadically throughout the United States and southern Canada. It has been designated as invasive, so care should be taken in growing this plant.

APRIL–JULY (97-4-6)
H-5A/MOS/URY
[○/D/✪]

Rosaceae

FLOWERING RASPBERRY; THIMBLEBERRY

Rubus odoratus Linnaeus

This shrub with branched, bristly, but not thorny canes 1–1.5 m tall has large, palmately lobed leaves 1–3 dm long. The attractive, 2–5 cm broad flowers are followed by edible, but not very flavorful, red raspberries that are quickly harvested by wildlife.

Thimbleberry is native to northeastern North America and extends into our area only in the mountains, where it often grows in colonies or clumps on shady road banks, in woodland borders, and along streams. In the right habitat, this attractive shrub would be a good choice for the wild garden. Native Americans used Thimbleberry for both food and medicine.

JUNE–AUGUST (97-5-1)
S-5A/SRL/URR
[◗/M/$/✪]

Rosaceae

BLACKBERRY

Rubus argutus Link

One of the several white-flowered, thorny species of *Rubus* found in North Carolina, with canes up to 2 m tall, this perennial often propagates by rhizomes, or runners, to form dense bramble thickets. The flowers are 4–6 cm across; the juicy, sweet black fruits are excellent when fully ripe.

A southeastern species north to Pennsylvania and Illinois, this Blackberry is often a weed in meadows, pastures, old fields, and woodland borders and along roadsides throughout North Carolina. There are at least a half dozen species of *Rubus* in our state that have highly edible fruits. The Cherokee used this species for food and medicine.

APRIL–MAY (97-5-9)
S-5A/POS/RRW
[❍/D/✪]

Rosaceae

FALSE VIOLET; ROBIN RUNAWAY

Dalibarda repens Linnaeus

This low (2–8 cm) perennial with pubescent, creeping stems has cordate, evergreen leaves, 3–5 cm long, that resemble those of some violets, which accounts for one of its common names. The flowers are about 1 cm across.

North Carolina is the southern limit of this northeastern species, our rarest member of the Rose family. It is known from mossy bogs in only three mountain counties and is state listed; do not dig! ✗

JUNE–SEPTEMBER (97-6-1)
H-5A/SCS/SRW
[◗/W/$/✪/*E*]

Rosaceae

SPREADING AVENS

Geum radiatum Gray

This pubescent native perennial, 2–5 dm tall, is one of seven yellow- , white- , or pink-flowered species of *Geum* in North Carolina and, with 1.3–2 cm long petals, has the largest flowers. The basal leaves are orbicular to obovate, 7–15 cm wide; the 2 to 5 stem leaves much smaller.

These rare native perennials are known only from high balds in seven of our mountain counties and a few adjacent areas of Tennessee; another yellow-flowered species, *G. virginianum*, is more widespread. Spreading Avens has both a state and federal listing. ✗

JUNE–AUGUST (97-7-7)
H-5A/SRS/URY
[◌/N/*E-SC*]

Rosaceae

HARVEST LICE; SMALL-FLOWERED AGRIMONY

Agrimonia parviflora Aiton

Stems of these densely pubescent perennials are 7–8 dm tall and usually clumped from a knotty, black rootstock. The 7 to 16 pinnately compound leaves per stem have 11 to 19 unequal, lanceolate leaflets that are 3–9 cm long. The small yellow flowers, in terminal racemes, are followed by the characteristic, rounded fruits with hooked bristles for ready transport on fur, feathers, and clothing—the harvest lice.

Found in marshes and alluvial habitats throughout our mountains and piedmont, this Cocklebur ranges throughout much of the eastern and central United States. A whole-plant herbal tea made from this plant has traditionally been used as an astringent for wounds and a treatment for jaundice and gout, among other conditions.

JULY–SEPTEMBER (97-9-2)
H-5A/PES/RRY
[◗/W/✪]

Rosaceae

CANADA BURNET

Sanguisorba canadensis Linnaeus

Although the small flowers of Canada Burnet have no petals, the showy stamens of the densely flowered, 6–25 cm long spikes are quite distinctive on the 9–15 dm tall stems. The large, basal, pinnate leaves have 7 to 15 serrate leaflets.

These native perennials get their name from the extensive range in Canada but make their way south along the mountains to their southern limit in North Carolina and Georgia. Canada Burnet is rare in bogs and seeps at high elevations in nine of our mountain counties. ✗

JULY–SEPTEMBER (97-10-3)
H-4B/PES/IRW
[❍/W/R]

Rosaceae

SWAMP ROSE

Rosa palustris Marshall

This much-branched, rhizomatous shrub, up to 2 m tall, has thorny, upright canes with pinnately compound leaves. The 5 to 9 elliptic leaflets are 2–6 cm long. The petals of these attractive flowers are 2–3 cm long.

A native of eastern North America, Swamp Rose is scattered but relatively frequent along the margins of streams, ponds, or swamp forests, essentially throughout our state. The very similar *R. carolina* is found in drier upland pastures and woodlands throughout the state.

MAY–JULY (97-11-1)
S-5A/POS/URR
[❍/M/$/✪]

Rosaceae

INDIAN PHYSIC; FAWN'S BREATH

Porteranthus trifoliatus (L.) Britton
[*Gillenia trifoliata* (L.) Moench]

The trifoliate leaves with serrate, oblanceolate leaflets, 4–8 cm long, and the 5 linear, slightly twisted, unequal petals (the longest 1–1.5 cm) help identify this open, branching perennial that grows to a height of 4–6 dm.

A native of northeastern North America, Indian Physic grows in the rich woods of the North Carolina mountains and northern piedmont and is an attractive plant for the wild garden. Indian Physic, as the name implies, was used for medicine, particularly as a purgative.

APRIL–JUNE (97-13-1)
H-5A/POS/PCW
[◗/M/$/✪]

Rosaceae

GOAT'S BEARD; BRIDE'S FEATHERS

Aruncus dioicus (Walt.) Fernald

As the species name implies, these tall perennials, 1–1.5 m high, are dioecious, having the male and female flowers on separate plants; the more attractive male plant is shown here. The large bipinnately or tripinnately compound leaves have ovate to lanceolate, serrate leaflets, 4–12 cm long.

A species of the eastern and central United States, Goat's Beard grows in the rich, moist woods of our mountains and piedmont. These plants make an attractive and conspicuous addition to the wild garden. Native Americans made use of the roots of Goat's Beard in the treatment of bee stings, swollen feet, and stomach problems.

MAY–JUNE (97-14-1)
H-5A/BOS/PRW
[◗/M/$/✪]

Rosaceae

HARDHACK; SPIRAEA

Spiraea tomentosa Linnaeus

The leaves of these 2 m tall native perennial shrubs are simple, tomentose beneath, elliptic to ovate, and 4–7 cm long. The flowers at the top of the inflorescence open first.

Though occasionally found in wet meadows or low woodland borders at scattered localities over most of the western and northern portions of the state, these plants are more common northward into Canada. Spiraeas are popular for landscape use and are recommended for wetland mitigation. Native Americans used flower and leaf teas of Hardhack to treat diarrhea, dysentery, and morning sickness.

JULY–SEPTEMBER (97-15-2)
S-5A/SNS/PRR
[◗/M/$/✪]

Rosaceae

NINEBARK

Physocarpus opulifolius (L.) Maximowicz

This 1–3 m tall shrub has simple, ovate, often 3-lobed leaves, 4–7 cm long and wide. The brown to orangish bark peels into thin strips—or sheets on larger trunks—giving the impression that this plant has "nine lives." The flower clusters are 2.5–5 cm broad.

Ninebark grows along stream banks and on moist cliffs in our mountains and piedmont and occurs widely in eastern North America. These shrubs are cultivated for their attractive flowers, foliage, and fruit and make wonderful landscape plants. The inner bark has been used extensively as an analgesic and emetic.

MAY–JULY (97-16-1)
S-5A/SOS/URW
[◗/M/$/✪]

Rosaceae

CRABAPPLE

Malus angustifolia (Ait.) Michaux

This small native tree or shrub, up to 10 m tall, has alternate, elliptic leaves that are 2.5–8 cm long and 1–4 cm wide. The delicate pink petals of the fragrant flowers are 1.5–2.5 cm long and soon fade to almost white. The hard, green, extremely sour apples are 2.5–3.5 cm in diameter.

A southeastern species with landscape potential is found north to Pennsylvania and grows along fencerows and low woodland margins, chiefly in our coastal plain and mountains. The Cherokee dried crabapples for food; rich in pectin, they have long been used in making jelly.

APRIL–MAY (97-18-2)
T-5A/SES/URR
[◯/N/$]

Rosaceae

MOUNTAIN ASH

Sorbus americana Marshall

The large, dark green pinnate leaves and, in spring, the compact clusters of white flowers, 6–15 cm wide, identify this striking small (up to 10 m tall) tree. In the fall, the clusters of bright red fruits, 4–6 mm in diameter, and the colorful yellow or red leaves make identification easy.

This species of northeastern North America follows the mountains to North Carolina, where it grows around rock outcrops, balds, and the margins of spruce-fir forests at high elevations. This attractive tree is highly recommended for landscape use. Native Americans throughout the range of this species have used the inner bark and the fruit for medicine.

JUNE–JULY (97-19-1)
T-5A/PES/URW
[◯/N/$/✪]

Rosaceae

RED CHOKEBERRY

Sorbus arbutifolia (L.) Heynhold

This rhizomatous shrub, 1–2 m tall, often forming large colonies, has simple, alternate leaves, 4–10 cm long. The flowers are in clusters at or near the ends of the short branches, the petals 4–7 mm long. The bright red, applelike fruits are 6–9 mm in diameter.

This species of the eastern United States is frequent in bogs, savannas, and low woodlands throughout our state and is a good choice for the wild garden.

MARCH–MAY (97-19-2)
S-5A/SES/URW
[◗/M/$]

Rosaceae

HAWTHORN

Crataegus flabellata (Bosc) K. Koch

This tall shrub, up to 8 m in height, with serrate, roughly heart-shaped leaves 4–8 cm long, has clusters of 1.3–2 cm broad flowers and, in the fall, red fruits up to 1.5 cm in diameter. These plants are so variable and so poorly understood botanically that over a dozen species and varieties have been described from the plants that are here considered to be a single species.

Native to eastern and central North America, some 13 species of Hawthorns are found throughout North Carolina in woodlands, pastures, and thickets. *Crataegus* species have a long history of medical use and, if you do not care about the taxonomic confusion, these attractive shrubs or small trees should be considered for wild landscaping. Flowers and fruits of all *Crataegus* species in America, Asia, and Europe are well known for their use in traditional herbal heart medicine, and there is potential for their use in modern medicine.

MAY–JUNE (97-20-4)
S-5A/SOS/URW
[◗/N/$/✪]

Rosaceae

SHADBUSH; SERVICEBERRY

Amelanchier arborea (Michx. f.) Fernald

This attractive small tree or shrub, up to 10 m tall, has drooping racemes of white or pale pink flowers with 10–20 mm long petals that appear before the leaves. The elliptic leaves are 4–13 cm long. The small, reddish-purple, applelike fruits are 6–10 mm in diameter and are popular with wildlife.

Native to the eastern and central United States, these plants grow in open, often rocky woodlands, chiefly in our mountains and piedmont. Serviceberry was used by the Cherokee for food and medicine and is an attractive possibility for native plant landscaping.

MARCH–MAY (97-21-1)
T-5A/SES/URW
[❍/N/$/✪]

Rosaceae

FIRE CHERRY; PIN CHERRY

Prunus pensylvanica L. f.

These native shrubs or small trees, up to 10 m tall, have smooth bark and finely toothed, long-acuminate leaves, 6–15 cm long and 2.5–5 cm wide. Unlike those of Wild Cherry, the flowers of Fire Cherry are in umbel-like clusters and the 6–8 mm fruit is red.

Ranging throughout most of our northern states and Canada, Fire Cherry follows the mountains south to Georgia. As the name implies, it often comes up in clearings after fire in our mountain counties. Native Americans made use of this species for food, fiber, and medicine. Note that the wilted leaves, twigs and seeds are toxic when ingested!

APRIL–MAY (97-22-8)
T-5A/SNS/UCW
[❍/N/$/✪/◉]

Rosaceae

WILD CHERRY; BLACK CHERRY

Prunus serotina Ehrhart

The leaves of this native tree, up to 25 m tall, are mostly 6–12 cm long. The small white-petaled flowers, about 4 mm long, are in racemes, 8–15 cm long. The juicy dark purple or black fruits are 1 cm in diameter and may be sweet or bitter.

A species of eastern North America, Wild Cherry is found in low woodlands and along fencerows throughout North Carolina. Black Cherry fruits were early used for flavoring, and the tree's beautiful, dark, richly colored wood (a major colonial export) was made into furniture. The aromatic inner bark was used as a medicine. Note, however, that the bark, wilted leaves, and seeds are highly poisonous!

APRIL–MAY (97-22-11)
T-5A/SES/RRW
[◗/N/$/✪/◉]

Fabaceae

SENSITIVE BRIER

Mimosa microphylla Dryander [*Schrankia microphylla* (Smith) Macbride]

The 1–2 m long, slender stem of this sprawling, native perennial vine has numerous small, sharp prickles. The bipinnately compound leaves with pinnae 2–5 cm long close at the slightest touch. The globose cluster of many small pinkish-purple flowers is about 2 cm in diameter.

Sensitive Brier is a southeastern species found along clay roadsides and woodland margins in all provinces, but it is rare or absent from the outer coastal plain.

JUNE–SEPTEMBER (98-2-1)
V-5A/BLE/KZR
[❍/N]

Fabaceae

REDBUD; JUDAS TREE

Cercis canadensis Linnaeus

This small, fast-growing tree, up to 12 m tall, has black bark and simple, heart-shaped leaves, 5–12 cm wide, which appear after the period of profuse flowering. The numerous, small, zygomorphic flowers are soon followed by flat, oblong, brown pods, 6–10 cm long, with numerous seeds.

These native trees, of the eastern and central United States north to New York and Nebraska, grow in cutover woodlands and along fencerows, especially on sweet or basic soils, chiefly in the piedmont. These trees make beautiful landscape plants especially in combination with Flowering Dogwood. A highly astringent tea made from Redbud's inner bark was used by Native Americans to treat fever and congestion and was a folk remedy for leukemia.

MARCH–MAY (98-4-1)
T-5A/SCE/UZR
[◗/N/$/☉]

Fabaceae

PARTRIDGE PEA; SENNA

Chamaecrista fasciculata (Michx.) Greene
[*Cassia fasciculata* Michaux]

These herbaceous annuals are 1.5–6 dm tall and have sensitive pinnate leaves with 12 to 36 leaflets. The flowers, which are not as strongly zygomorphic as those of most legumes, have petals that are 1–2 cm long and differ in size.

A native of the eastern and central United States, these weedy plants frequently grow in old fields and along roadsides and forest margins throughout North Carolina. Senna can be grown easily from seed for wild landscaping. The Cherokee used a tea brewed from its roots to relieve fatigue. Note: the seed is mildly toxic if ingested.

JUNE–SEPTEMBER (98-5-5)
H-5A/PLE/UZY
[❍/N/✪/△]

Fabaceae

YELLOWWOOD

Cladrastis kentukea (Dum.-Cours.) Rudd
[*Cladrastis lutea* (Michx. f.) K. Koch]

The showy racemes of fragrant, white flowers of this native tree, up to 20 m tall, are 1–3 dm long. The alternate, pinnately compound leaves, 4–6 dm long, have 7 to 9 leaflets. The gray or light brown bark of the trunk is smooth; and the wood, which has been used as a dye source, is, as the common name indicates, yellow.

This tree, native to the central United States east to North Carolina, is found in the rich woods of only a few of our mountain counties but is often planted elsewhere as an ornamental.

APRIL–MAY (98-8-1)
T-5A/POE/PZW
[◗/M/$]

Fabaceae

HAIRY WILD INDIGO

Baptisia cinerea (Raf.) Fernald & Schubert

These tawny-pubescent, 3–6 dm tall perennials have coriaceious trifoliate leaves with obovate leaflets, 3–7 cm long. The bright yellow, 2–2.5 cm long flowers are usually in solitary, terminal racemes.

Ranging from Virginia to Georgia, this Wild Indigo grows in sandhills and sandy woods in our coastal plain and lower piedmont. These attractive perennials make a good choice for drier sites in the wild garden. All parts of the plant are mildly poisonous.

MAY–JUNE (98-9-3)
H-5A/TBE/RZY
[❍/D/$/△]

Fabaceae

YELLOW WILD INDIGO

Baptisia tinctoria (L.) R. Brown

These glabrous, much-branched perennials, 3–10 dm tall, have trifoliate leaves with obovate leaflets that blacken when dry. The numerous racemes have smaller flowers (9–16 mm) than other *Baptisia* species. The subglobose pods, 7–9 mm long and 6–8 wide, are almost black at maturity.

Yellow Wild Indigo is found in open woods and clearings throughout our state and through much of the eastern United States. While of interest for its form in the wild flower garden, it is not as colorful and eye-catching as many other of our native legumes. The roots of this species have a long history of Native American and folk medicinal use as an emetic, purgative, astringent, and antiseptic, and the plant has also been used as a blue dye.

APRIL–AUGUST (98-9-7)
H-5A/TNE/RZY
[◗/D/$/✪/△]

Fabaceae

WHITE WILD INDIGO

Baptisia alba (L.) R. Brown

These often robust perennials, 5–15 dm tall, have trifoliate leaves with elliptic leaflets, 2–4 cm long. The flowers are 1.2–1.8 cm long, and the cylindric pods are 2–3.5 cm long.

These attractive plants are found in open woods and clearings in a few counties of our piedmont and coastal plain and more commonly south to Florida and Alabama. Baptisia species make excellent additions to the wild flower garden as they are attractive in bud, flower, fruit, and leaf and, once established, are hardy and need little care. All *Baptisia* species are mildly toxic.

MAY–JULY (98-9-9)
H-5A/TEE/RZW
[◗/D/$/✪/△]

Fabaceae

BUSH PEA

Thermopsis villosa (Walt.) Fernald & Schubert

A close relative of *Baptisia*, these native, rhizomatous perennials, 0.6–1.6 cm tall, have racemes 1–3 dm long. The leaflets of the trifoliate leaves are 4–10 cm long. The pods, or legumes, of this species are villous, or densely pubescent with long hairs.

These plants of the northeastern states follow the mountains south to Georgia. They are infrequent in clearings or along forest margins in a few of our western counties and are a good candidate for the sunny garden border.

MAY–JUNE (98-10-1)
H-5A/TEE/RZY
[○/N/$]

Fabaceae

WILD LUPINE; SUNDIAL LUPINE

Lupinus perennis L.

These attractive perennials, 2–6 dm tall, have villous, palmately compound leaves of 7 to 11 leaflets. The 1.5 cm long flowers are in terminal racemes and are usually blue but may rarely be pink or white.

Wild Lupine is found in our sandhills and in clearings and open woods of the coastal plain and more widely through much of the eastern United States. These eastern relatives of the more prolific Texas Bluebonnet can be grown from seed and make attractive additions to drier areas of the wild flower garden. Note: the seeds are poisonous if ingested.

APRIL–MAY (98-12-1)
H-5A/MBE/RZB
[○/D/$/⊕/△]

Fabaceae

BUFFALO CLOVER

Trifolium reflexum Linnaeus

This villous, annual or biennial clover, 2–5 dm tall, has trifoliate leaves with ovate to elliptic leaflets, 1.3–5 cm long. The globose heads, 2–2.5 cm in diameter, contain 8–12 mm long flowers with red or roseate standards exceeding the white to pinkish wing and keel, all becoming dark brown.

Native to the south and central United States north to Pennsylvania and west to Nebraska, Buffalo Clover grows in open woods, in clearings, and along roadsides in all three of North Carolina's provinces. Many of our 14 species of clovers are European introductions. All are important forage for wildlife.

APRIL–AUGUST (98-14-10)
H-5A/TOE/KZR
[◗/N]

Fabaceae

YELLOW SWEET CLOVER

Melilotus officinalis (L.) Lam.

This erect or sprawling introduced biennial weed, 0.4–2 m tall, has trifoliate, cloverlike leaves, 1–2.5 cm long. The showy yellow racemes, 4–12 cm long, have flowers up to 1 cm in length.

Easily spotted along roadsides and in waste areas throughout our state, and now throughout much of North America, species of *Melilotus* are sometimes planted as a forage crop for cattle or as a source of nectar for honey. They are not, however, appropriate for the garden except to plow under for green manure. The dried flowering plants of Yellow Sweet Clover have had a long use in folk medicine—as a tea, poulticed, and smoked—for a variety of ailments.

APRIL–OCTOBER (98-15-2)
H-5A/TBE/RZY
[❍/N/✪]

Fabaceae

FALSE INDIGO; LEAD PLANT

Amorpha fruticosa Linnaeus

This native, branching shrub, 1.5–4 m tall, has alternate, pinnately compound leaves, 1–3 dm long, with 11 to 25 ovate to elliptic leaflets. The numerous erect racemes, 1–2 dm long, have many small purple flowers with bright orange anthers.

These plants range over much of temperate North America and grow along stream banks and in open woods at scattered localities in all provinces of North Carolina. Lead Plant is often cultivated since it is a hardy and attractive plant for a variety of moisture conditions. This species was used by western Native American tribes for fiber and arrows.

APRIL–JUNE (98-18-5)
S-5A/POE/RZB
[◗/N/$]

Fabaceae

PENCIL FLOWER

Stylosanthes biflora

This prostrate to erect, 1–5 dm long perennial has trifoliate leaves with elliptic to oblanceolate leaflets, 1.5–4 cm long. The orange, yellow, or rarely white flowers have standards 5–9 mm long and, though small, are quite attractive.

These plants range from New York to Texas in open woods and dry barrens and are found throughout our state. Their tolerance of dry soil makes them good plants for the wild garden.

JUNE–AUGUST (98-25-1)
H-5A/TEE/IZY
[◗/D/✪]

Fabaceae

BEGGAR'S TICKS; NAKED-FLOWER TICKTREFOIL

Desmodium nudiflorum (L.) DC

Few hikers know the small, pink, papilionaceous flowers of this perennial with a slender, arching, flowering stem 3–10 dm long and a sterile shoot up to 3 dm long bearing 4 to 7 trifoliate leaves. Most people, however, are quite familiar with the flattened, triangular segments of the fruit that adhere to clothing, fur, and feathers in the late summer and fall. These modified, indehiscent legumes with 2 to 4 segments are covered with minute hooked hairs that are responsible for their attachment and are said to have been the model for the invention of Velcro.

Widespread in woodlands throughout the state and in most of eastern North America, *D. nudiflorum* and the 22 other species of Beggar's Ticks found in North Carolina will all be recognized by their distinctive "sticky" fruits. None is a likely choice for the wildflower garden. The Cherokee chewed the roots to reduce gum inflammation.

JULY–AUGUST (98-26-1)
H-5A/TEE/RZR
[◗/N/✪]

Fabaceae

ROUND-HEADED BUSH CLOVER; LESPEDEZA

Lespedeza procumbens Michaux

The slender, pubescent, trailing or weakly ascendent stems of this perennial vine, 3–12 dm long, have alternate, pubescent, trifoliate leaves with ovate leaflets, 0.8–2.5 cm long. Two forms of flowers are found within the racemes of this and most Lespedeza species: cleistogamous, apetalous, usually sessile flowers that do not open but self-pollinate, and petaloid, purple, chasmogamous flowers with the typical papilionaceous form.

These plants are found in fields and open, dry woods as well as on roadsides throughout our state and the eastern United States to New Hampshire and Michigan.

JULY–SEPTEMBER (98-27-3)
H-5A/TOE/RZR
[○/D]

Fabaceae

AMERICAN WISTERIA

Wisteria frutescens (L.) Poir.

This attractive native, high-climbing, woody vine, 2–15 m long, is related to the widely planted Asian species of Wisteria but is not as strong or aggressive and usually will not harm the trees on which it climbs. The leaves, 1–3 dm long, have 9 to 15 leaflets. Also unlike its Asian counterpart, American Wisteria has a glabrous legume.

Growing primarily along the borders of lowland woods and streams of the coastal plain, this native ranges throughout the eastern United States north to New York. Relatively easy to grow, this Wisteria makes a great addition to the wild garden. Warning: the seeds are toxic if ingested.

APRIL–MAY (98-31-1)
V-5A/POE/RZB
[◗/N/$/△]

Fabaceae

BLACK LOCUST

Robinia pseudo-acacia Linnaeus

Widely planted as ornamentals, these thorny, native trees, up to 25 m tall, have attractive, drooping racemes of fragrant flowers, 1–2 dm long. The alternate, pinnate leaves are 2–3 dm long, with 7–19 elliptic leaflets. The flat legumes are 5–10 cm long.

A native of the southeastern United States and now spread throughout much of temperate North America, this somewhat weedy tree grows in old fields or along roadsides, fencerows, and woodland margins, chiefly in our piedmont and mountains. The hard wood, once used as the hub of wagon wheels, today makes durable fence posts and pilings. Black Locust bark has been used in both Native American and folk medicine as an emetic and purgative; all parts of these trees are toxic if ingested.

APRIL–JUNE (98-32-1)
T-5A/PEE/RZW
[○/N/$/✪/△]

Fabaceae

BRISTLY LOCUST

Robinia hispida Linnaeus

This attractive, native, rhizomatous shrub, 1–2 m tall, often forms large colonies. The pinnately compound leaves are 1.5–3 dm long, with 7 to 13 leaflets, 1.5–5 cm long, and the attractive, inodorous flowers, 2–3 cm long, are in racemes up to 13 cm long.

Bristly Locust ranges broadly through much of the eastern and central United States and is found along road banks and woodland margins at scattered localities in all provinces of North Carolina. The Cherokee used this plant for medicine, building materials, and weapons. Bristly locust makes a colorful addition to the wild garden.

MAY–JUNE (98-32-4)
S-5A/PEE/RZB
[◗/N/$/✪]

Fabaceae

GOAT'S RUE

Tephrosia virginiana (L.) Persoon

This low, villous native perennial has unbranched stems, 2–7 dm tall, terminating in compact racemes, 4–8 cm long. The pinnately compound leaves have 15 to 25 elliptic to linear leaflets, 1–3 cm long. The large bicolored flowers are 1.5–2 cm long; the large, white, upper petal is the "standard"; the pink lateral petals are the "wings"; and the two fused lower petals are the "keel" of this typical Pea flower.

A relatively frequent plant along dry roadsides, clearings, and open woods more or less throughout the state and much of the eastern and central United States, Goat's Rue is a good candidate for a full-sun garden. Extensive medicinal use has been made of this species, and it has been used as a fish poison and insecticide; all parts are mildly toxic if ingested.

MAY–JUNE (98-34-1)
H-5A/PEE/RZR
[❍/D/$/✪/⚠]

Fabaceae

CANADIAN MILKVETCH

Astragalus canadensis Linnaeus

The erect, glabrous stems of this 5–15 dm tall perennial have pinnate leaves, 1–2 dm long, with 13 to 31 oblong to elliptic leaves, 1–4 cm long. The somewhat reflexed, pale, 12–15 mm long flowers are in long-peduncled, axillary racemes; the numerous ovoid, 2-celled legumes are 1–2 cm long.

This widespread native species found throughout much of the United States and southern Canada grows in rich woods and along stream banks in our mountains and upper piedmont. Throughout its range, Native Americans have used the roots of Canadian Milkvetch for both medicine and food. Seed is available commercially for growing this interesting Legume.

JUNE–AUGUST (98-35-2)
V-5A/PEE/RZW
[◗/N/$/✪]

Fabaceae

CAROLINA VETCH

Vicia caroliniana Walter

The few to numerous white flowers of this slender, trailing to erect, native, perennial, 3–10 dm vine are borne on axillary racemes. The pinnately compound leaves consist of three or more pairs of linear-oblong leaflets, 1–2 cm long, with the terminal leaflet modified into a tendril. Racemes are 6–10 cm long with 7 to 20 flowers, each about 1 cm long.

Carolina Vetch grows in open woods and along woodland margins throughout the eastern United States north to New York, but is primarily found in our mountains and piedmont. The Cherokee made extensive use of this species to alleviate pain and rheumatism. It is a docile plant for the wild garden.

APRIL–JUNE (98-36-12)
V-5A/PEE/RZW
[◗/N/✪]

Fabaceae

SPURRED BUTTERFLY PEA

Centrosema virginianum (L.) Bentham

This trailing or climbing, pubescent perennial vine is 5–15 dm long and has trifoliolate leaves, each leaflet 2–7 cm long. The conspicuous flowers, in axillary racemes, have a 2.5–3.5 cm long standard that is spurred. The flattened, sessile legumes are 7–14 cm long.

Ranging throughout the southeastern and central states north to New Jersey and Illinois, Spurred Butterfly Pea is found in open woods and clearings throughout our state. This attractive Pea has potential for drier sites in the wild garden.

JUNE–AUGUST (98-41-1)
V-5A/TOE/RZB
[◗/D]

Fabaceae

BUTTERFLY PEA

Clitoria mariana Linnaeus

This slender, twining or trailing native perennial vine is 5–10 dm long and has trifoliolate leaves, each leaflet 2–7 cm long. The few-flowered axillary racemes have large, conspicuous flowers, with a standard 4–6 cm long and 3–4 cm wide that is not spurred. The legume is 3–6 cm long.

Ranging throughout the Southeast north to New York, Butterfly Pea is found infrequently, but essentially throughout North Carolina, growing in open, usually dry woods and clearings. Along with the similar Spurred Butterfly Pea, this species has potential for drier sites in the wild garden.

JUNE–AUGUST (98-42-1)
V-5A/TOE/RZB
[◗/D/✪]

Fabaceae

WILD BEAN

Strophostyles umbellata (Muhl.) Britton

This trailing or weakly twining perennial vine, up to 2 m long, has trifoliolate leaves, each narrowly to widely ovate leaflet 2–5 cm long. The 1–2.5 dm long racemes have few to several clusters of pink or pale purple flowers, the standards 1–1.4 cm long. The legume is 3–6.5 cm long and 4 mm broad.

Ranging throughout the southeastern and central states north to New York, Wild Bean is found throughout our area in fields, woods, and clearings.

JUNE–SEPTEMBER (98-46-2)
V-5A/TOE/RZR
[◗/N]

Oxalidaceae

MOUNTAIN WOOD SORREL; WOOD SHAMROCK

Oxalis montana Rafinesque
[*Oxalis acetosella* Linnaeus]

The leaves and the 6–15 cm tall flowering stems of this native perennial arise directly from a creeping rhizome. The heart-shaped leaflets are 1–2 cm long, and the single chasmogamous flower, with petals 1–1.5 cm long, overtops the leaves. The shorter cleistogamous flowers lie under the leaves.

North Carolina is near the southern limit for these plants, which grow in cool moist forests in eastern Canada and the northeastern United States. Often under hemlock or spruce trees, this Wood Sorrel is found at high elevations in our western counties. An attractive addition for the wild garden (if it can stand the heat at lower elevations), this species was used by Native Americans for food and as a dye; but all parts of the plant are mildly toxic.

MAY–SEPTEMBER (100-1-1)
H-5B/TBE/SCW
[●/M/$/△]

Oxalidaceae

PINK WOOD SORREL

Oxalis violacea Linnaeus

Each leafless flower stalk of these low, 1–2 dm tall perennials bears umbellate clusters of pink (not violet) chasmogamous flowers with petals 10–18 mm long. Leaves and flowers arise from small bulbs and are rhizomatous, as is the case with *O. montana*.

Native to much of the eastern and central United States, this Wood Sorrel is found in moist to dry woodlands at numerous scattered localities across North Carolina. Propagation by seed should be successful for this attractive perennial. Native Americans used Pink Wood Sorrel for both food and medicine.

APRIL–MAY (100-1-3)
H-5B/TBE/UCR
[◗/N/$/✪/△]

Oxalidaceae

SOURGRASS; OXALIS

Oxalis grandis Small

This is the largest of about half a dozen yellow-flowered species of *Oxalis* in North Carolina. This annual native is 3–10 dm tall, and the leaflets, edged with a thin maroon line, are 2–5 cm wide. The cymose inflorescence, with flowers 14–18 mm long, surpasses the leaves.

This species of the Mid-Atlantic states is found in North Carolina in the rich woods of the mountains and, less frequently, the upper piedmont.

MAY–JUNE (100-1-8)
H-5B/TBE/UCY
[◗/M/$/△]

Geraniaceae

WILD GERANIUM

Geranium maculatum Linnaeus

This erect, rhizomatous perennial grows to a height of 3–7 dm. Stem leaves are 4–10 cm long and 6–16 cm wide and are cleft into 5 to 7 segments. The flowers are 2.5–4 cm across in few-flowered, umbellate clusters.

Native to the eastern and central United States, these plants grow in moist woodlands and in rich coves, chiefly in our mountains and piedmont. They can be spectacular in the wild garden, where plants do well in light shade or partial sun. Eastern and central Native Americans used Wild Geranium roots as an astringent for oral health and as an antidiarrheal.

APRIL–JUNE (101-1-1)
H-5A/SRC/URR
[◗/M/$/✪]

Geraniaceae

CAROLINA CRANESBILL

Geranium carolinianum Linnaeus

This erect, pubescent winter annual or biennial, up to 6 dm tall, has palmately dissected basal and cauline leaves, 1.5–3 cm long. The paired pale pink flowers, with 4–6 mm long petals, are followed by the distinctive beaked fruits that account for the common name.

Native throughout most of the United States and southern Canada, Carolina Cranesbill is a common weed of disturbed habitats, gardens, fields, pastures, and roadsides throughout our state.

MARCH–JUNE (101-1-4)
H-5A/SOC/URR
[❍/N]

Polygalaceae

FRINGED POLYGALA; GAY WINGS

Polygala paucifolia Willdenow

The few elliptic to oval leaves, 1.5–4 cm long, of this low, 8–15 cm tall, rhizomatous perennial are crowded near the top of the stem. The 1 to 5 colorful, chasmogamous flowers, the largest found in any of North Carolina's Polygalas, are 13–19 mm long. Cleistogamous flowers are found on separate, usually nonleafy stems.

This plant of northeastern North America ranges through the mountains to Georgia and grows in deciduous forests at high elevations in a few western North Carolina counties. This attractive plant is highly recommended for the wild shade garden. The Iroquois used infusions of the whole plant and leaf poultices to treat abscesses and sores.

APRIL–JUNE (106-1-1)
H-3A/SEE/UZR
[◗/M/$/✪]

Polygalaceae

POLYGALA

Polygala curtissii Gray

This slender, branched, native annual, 1–4 dm tall, has linear leaves, 1–2 cm long. The numerous small flowers, each 3–5 mm long, are borne in compact racemes at the ends of short terminal branches that continue to develop and produce buds over a long season.

Found from Pennsylvania to Georgia, these plants grow in old fields and along moist woodland borders at many scattered localities in the piedmont and mountains of North Carolina.

JUNE–OCTOBER (106-1-7)
H-3A/SLE/RZR
[◯/N]

Polygalaceae

ORANGE MILKWORT

Polygala lutea Linnaeus

This native biennial, 1–4 dm tall, is the most colorful species of *Polygala* found in North Carolina. The compact racemes, 1–3.5 cm long and 1–2 cm in diameter, continue to grow and produce flower buds throughout the summer; the small fruits fall when ripe, leaving scars on the flower stalk beneath the existing blooms.

A southeastern coastal species from New York to Louisiana, Orange Milkwort is fairly common on roadsides and in savannas and pine barrens throughout our coastal plain. This colorful species is a good choice for the sunny wild garden. Records include medicinal use for ailments of the heart and blood by both the Choctaw and Seminole.

APRIL–OCTOBER (106-1-14)
H-3B/SBE/RZY
[◯/M/✪]

Euphorbiaceae

STINGING NETTLE

Cnidoscolus stimulosus (Michx.) Engelmann & Gray

These low, rhizomatous, monoecious perennials, 0.5–10 dm tall, with alternate, lobed leaves are armed with numerous stinging hairs. The fragrant flowers have no petals, but the calyx is petaloid and fused into a tube 1–1.5 cm long.

A native of the southeastern coastal plain, these plants are relatively frequent in the sandhills and old sandy fields of eastern North Carolina. The stinging hairs can inflict pain if touched and should be avoided.

MARCH–AUGUST (107-1-1)
S-5A/SLE/RAW
[❍/D/☞]

Euphorbiaceae

CROTON

Croton punctatus Jacquin

This rather coarse, branched, monoecious, native annual or short-lived perennial is up to 1 m tall and is profusely dotted with small scales and glands that give the entire plant a brownish-gray cast. The ovate leaves are 2–6 cm long and 1–4 cm wide. The small, inconspicuous female flowers produce 3-lobed capsules that may attain a width of 5–7 mm.

These plants grow only on, or just behind, the sand dunes along the southeastern coast, from Texas to their northern limit in Dare County, North Carolina.

MAY–NOVEMBER (107-2-2)
H-VA/SOE/RRW
[❍/D]

Euphorbiaceae

POINSETTIA; PAINTED LEAF

Euphorbia heterophylla Linnaeus

A wild relative of the colorful, cultivated Poinsettias, these annuals, up to 1 m tall, have upper leaves 1–4 cm wide just below the inflorescence of small inconspicuous flowers. The leaves vary in color, pattern, and shape.

A semiweedy plant of the Gulf and South Atlantic coastal plains as far north as Virginia, the species is known from disturbed habitats in seven of our eastern counties and should grow easily from seed.

JUNE–OCTOBER (107-11-2)
H-NA/SEL/KAY
[❍/D]

Euphorbiaceae

CAROLINA IPECAC

Euphorbia ipecacuanhae Linnaeus

These low, exceptionally variable, monoecious, deep-rooted perennials, up to 3 dm tall, fork repeatedly and usually form a green mat of vegetation. The opposite, fleshy leaves, 1–7 cm long, may be red or green and are highly variable in shape, from linear to oblanceolate. The minute female flowers produce small, 3-lobed capsules characteristic of the Spurge family.

These perennials grow in the deep sandy soil of pinelands and turkey-oak woods of our coastal plain and in the coastal states from Connecticut to Georgia. These plants were widely used medicinally as an emetic.

MARCH–MAY (107-11-10)
H-NA/SEE/KAY
[❍/D/✪]

Euphorbiaceae

MILKWEED; FLOWERING SPURGE

Euphorbia corollata Linnaeus

The sap of these slender, branched, 1–2 dm tall perennials is milky, which is a characteristic of most members of this family. The numerous flowers, about 5 mm broad, lack petals and sepals but have 5 showy, white, petaloid glands that attract pollinators.

This widespread species of the eastern and central United States grows in old fields, clearings, and waste places throughout most of North Carolina except the outer coastal plain. A Milkweed root tea was used as both a laxative and emetic but is now considered mildly toxic when ingested. The white sap is a skin irritant.

MAY–SEPTEMBER (107-11-13)
H-NA/SLE/URW
[○/D/✪/⚠☞]

Anacardiaceae

POISON SUMAC; THUNDERWOOD

Toxicodendron vernix (L.) Kuntze
[*Rhus vernix* Linnaeus]

This shrub or small tree, up to 5 m tall, is identified by its smooth, light gray bark and pinnately compound leaves with 7 to 13 leaflets, 5–12 cm long and 2–5 cm wide, which are especially colorful in the fall. But do not touch them; they are poisonous! The small green flowers are in an axillary inflorescence and produce small, globose white fruits.

Native to the eastern United States, this shrub grows in bogs, swamp margins, bays, and wet pine barrens of our mountains and coastal plains. Be careful: all parts of this shrub, which is closely related to Poison Oak and Poison Ivy, produce a contact skin poison that can cause an extremely severe reaction.

MAY–JUNE (110-1-1)
S-5A/PEE/PAW
[◗/W/✪/☞]

Anacardiaceae

POISON IVY

Toxicodendron radicans (L.) Kuntze
[*Rhus radicans* Linnaeus]

The 3 smooth, acuminate leaflets of Poison Ivy, each 5–12 cm long, are easy to recognize. The 3 leaflets of the related Poison Oak, *T. pubescens*, usually a low shrub and not a vine, are lobed and more rounded—but equally toxic! Panicles of small flowers occur in the leaf axils. The small white berries, 4–5 mm broad, are eaten (and widely spread) by many different birds.

Native over the eastern United States west to Texas, this common perennial, with beautiful fall color, grows by roadsides, along fencerows, in low woodlands, and in waste places throughout the state. Warning: Contact, either direct or indirect (even from pets or smoke produced by burning vines!), with any part of these poisonous, woody vines results in severe inflammation and blistering of the skin for many people.

APRIL–MAY (110-1-2)
V-5A/PES/PAW
[◗/N/✪/☞]

Anacardiaceae

SMOOTH SUMAC

Rhus glabra Linnaeus

This nontoxic shrub or small tree, up to 6 m tall, has pinnate leaves with 11–31 lanceolate leaflets, 5–10 cm long. The dense inflorescence is often 2 dm long and is quite conspicuous in fruit.

Widespread throughout temperate North America, Smooth Sumac grows in dry soils of old fields, along woodland borders, and in waste places in our mountains, piedmont, and inner coastal plain. Fruits are often used in dried arrangements or can be soaked in cool water to produce a tasty "lemonade." Both the attractive fall foliage and fruit make this plant popular for landscape use. The leaves, bark, and fruit of Smooth Sumac were used by dozens of Native American tribes as a treatment for asthma, fevers, and dysentery, among many ailments.

MAY–JULY (110-1-8)
S-5A/PNS/PAW
[❍/D/$/✪]

Cyrillaceae

LEATHERWOOD; TITI

Cyrilla racemiflora L.

These large shrubs or small trees, up to 8 m tall, have simple, alternate semi-evergreen, obovate leaves, up to 10 cm long, and numerous slender, terminal racemes, 5–15 cm long, of small, white flowers.

Common in pocosins, bay forests, and low pinelands of our coastal plain and throughout the southeastern coastal states from Virginia to Texas, Leatherwood makes a handsome landscape plant.

MAY–JULY (111-1-1)
S-5A/SBE/RRW
[◗/M/$]

Aquifoliaceae

AMERICAN HOLLY

Ilex opaca Aiton

This slow-growing tree, up to 15 m tall, with smooth, gray-green bark, has leathery, spiny evergreen leaves that are 5–10 cm long. The trees are dioecious, thus the colorful red berries, 8–10 mm in diameter, occur only on the trees that produce the small white female (pistillate) flowers.

These native trees grow scattered in low deciduous woods throughout North Carolina and the eastern United States to Texas and are widely used in landscape plantings—especially the female trees. Native Americans used the leaves and fruits to treat indigestion, colds, measles, and many other ailments, but the fruits are toxic if ingested.

APRIL–JUNE (112-1-1)
T-4A/SES/URW
[◗/N/$/✪/△]

Aquifoliaceae

YAUPON; BLACK DRINK

Ilex vomitoria Aiton

The thick, lustrous, elliptic evergreen leaves of this large shrub or, rarely, small tree, up to 8 m tall, are 1–5 cm long. As with most Hollies, both male and female flowers are located in the axils of the leaves and are relatively insignificant but are followed by brilliant red "berries" (actually drupes) that are quite attractive.

Yaupon is common in maritime forests of the coastal plain and throughout the southeastern coastal states to Texas. Its leaves contain caffeine and were dried and used as a ceremonial "black drink" (and as a trade item) by Native Americans. The stimulating drink was primarily a laxative or purgative (hence the species name). Yaupon is easy to grow, hardy, and quite attractive for landscape or garden use.

MARCH–MAY (112-1-3)
S-4A/SNS/URW
[◗/M/$/✪/△]

Aquifoliaceae

MOUNTAIN HOLLY

Ilex montana Torrey & Gray

This dioecious native shrub or small tree, up to 10 m tall, has small clusters of red berries, 5–9 mm in diameter, on the female plants. The thin, deciduous, elliptic leaves are 4–18 cm long.

These plants, native to the eastern states from New York to Alabama, grow in mixed hardwood and pine woodlands, chiefly in our mountains and piedmont. As with other Holly species, the seeds have a complex "double dormancy" system and are difficult to germinate, but hardwood cuttings can be rooted fairly easily.

APRIL–JUNE (112-1-6)
T-4A/SNS/URW
[◗/N/$]

Celastraceae

STRAWBERRY BUSH; HEARTS A'BUSTIN

Euonymus americanus Linnaeus

These slender, upright or trailing shrubs with smooth evergreen stems, grow up to 2 m tall. They have thick, opposite leaves, 3–9 cm long, and inconspicuous, flat, greenish-cream flowers with petals 2.5–4 mm wide. The warty red fruits, 1–2 cm in diameter, open in the early fall, exposing the crimson seeds.

Generally a southeastern species that occurs north to New York and west to Texas, these stoloniferous plants are relatively frequent in mixed deciduous forests and low woodlands throughout our area. They make attractive landscape plants, and physicians used the bark and seeds as a tonic, laxative, diuretic, and expectorant. Note, however, that the fruit, seeds, and bark are poisonous when ingested!

MAY–JUNE (113-3-2)
S-50/SNS/SRW
[◗/N/$/⊕/△]

Aceraceae

MOUNTAIN MAPLE

Acer spicatum Lamarck

This small, monoecious tree, up to 10 m tall, has slender panicles of greenish-white flowers in erect, terminal racemes. The opposite leaves are usually 3-lobed, sharply toothed, and up to 12 cm long and wide. The winged fruits are red when young but turn brown when ripe.

A native of the northeastern United States and Canada that follows the mountains south to Georgia, these trees grow in clumps in the rich, rocky woods at elevations above 4,000 feet in our mountains. Several tribes of Native Americans used the bark of Mountain Maple for an eye medicine.

MAY–JULY (115-1-2)
T-50/SRL/RRY
[●/M/$/✪]

Hippocastanaceae

RED BUCKEYE

Aesculus pavia Linnaeus

Easily recognized when in bloom by the large, open raceme of 2–3 cm long red flowers, this low native shrub (or rarely a small tree), 1–4 m tall, has palmately compound leaves with 5 slender leaflets 6–17 cm long.

These colorful natives of the southeast are occasionally found in the low woodlands and along swamp margins in our southern coastal plain counties. However, they have become a popular landscape plant and often hybridize with other (yellow-flowered) *Aesculus* species growing naturally nearby. The Cherokee used the seeds, bark, and roots for medicine, but, take care—the seeds and leaves of Red Buckeye are highly toxic!

APRIL–MAY (116-1-2)
S-40/PBS/RZR
[◗/M/$/✪/◉]

Hippocastanaceae

YELLOW BUCKEYE

Aesculus flava Ait.
[*Aesculus octandra* Marshall]

This handsome tree, up to 30 m tall, has opposite, palmately compound leaves. Its large racemes of 1.6–2.5 cm long, pale yellow flowers, attract bumblebees, which are the primary pollinators. The large, rounded, brown, leathery fruits, 5–8 cm in diameter, contain 1 to 4 brown seeds, or "Buckeyes." Another Buckeye common in the piedmont, *A. sylvatica*, has similar flowers and fruit but is a shrub 1–3 m high.

Yellow Buckeye is more common northward, ranging from Pennsylvania to Georgia and inland to Illinois. In North Carolina it occurs in the rich cove forests and deciduous woodlands of our mountains. Warning: buckeyes are poisonous and were once crushed and thrown into quiet streams to stun or kill the fish for easy harvest.

APRIL–JUNE (116-1-3)
S-40/MBS/RZY
[◗/M/$/◉]

Balsaminaceae

PALE JEWELWEED; TOUCH-ME-NOT

Impatiens pallida Nuttall

This branched, succulent-stemmed native annual herb is 5–15 dm tall, with thin, elliptic leaves, 5–15 cm long. The spurred, zygomorphic yellow flowers, usually without spots or with inconspicuous spots, are 2–3 cm long. The green capsule is "spring loaded" and explodes when touched, throwing the seeds up to 3 m—thus the common name, Touch-me-not. The name Jewelweed presumably comes from the bright blue inner seed coat.

A plant of northeastern North America south to Georgia and Oklahoma, Pale Jewelweed grows in moist areas with neutral or basic soil in the mountains and a few piedmont counties in North Carolina. Crushed leaves are recommended as an antidote to Poison Ivy exposure. The seeds are easy to collect (just before the fruits pop!) and are easy to grow.

JULY–SEPTEMBER (118-1-1)
H-5A/SED/RZY
[◗/M/$/⊙]

Balsaminaceae

SPOTTED JEWELWEED; TOUCH-ME-NOT

Impatiens capensis Meerburgh

This native annual herb is very similar to Pale Jewelweed except that its deep orange to red flowers are clearly spotted and are attractive to bumblebees, butterflies, and hummingbirds.

Unlike Pale Jewelweed, this plant grows in acid soils in moist open areas along streams and in low woods throughout most of North Carolina and eastern and central North America. Records of Native American use as a medicine and dye are extensive for this species throughout its range. The crushed leaves of this species, as well as *I. pallida*, are thought to counter the effects of contact with Poison Ivy.

MAY–FROST (118-1-2)
H-5A/SED/RZY
[❍/M/$/✪]

Rhamnaceae

NEW JERSEY TEA

Ceanothus americanus Linnaeus

A low, often freely branched shrub, up to 1 m tall, this plant has terminal clusters of numerous small flowers. The leaves are alternate, mostly 3–8 cm long, and pubescent to pilose beneath.

These shrubs are relatively frequent in open woodlands, on forest margins, and along roadsides throughout our state and much of eastern and central North America. The leaves at one time were dried and used as a tea substitute, and there are extensive records of the use of roots in medicine by Native Americans throughout the range of the species.

MAY–JUNE (119-2-1)
S-5A/SNS/PRW
[○/D/$/✪]

Vitaceae

VIRGINIA CREEPER; WOODVINE

Parthenocissus quinquefolia (L.) Planchon

This often cultivated, high-climbing woody vine is a member of the Grape family but is often confused with Poison Ivy. The five 6–12 cm long leaflets, however, identify it as Virginia Creeper. Panicles of small greenish-yellow flowers are terminal or in the upper leaf axils and thus rarely seen; the dark blue fruits, 5–9 mm in diameter, also distinguish this vine from Poison Ivy.

A native of the eastern and central United States, Virginia Creeper is common in dry or rich woods, on sand dunes, and often in waste places throughout the state. Both roots and leaves have been used for medicine but note that the berries are highly toxic!

MAY–JULY (120-1-1)
V-5A/PES/PRW
[◗/D/$/✪/◉]

Vitaceae

MUSCADINE; SCUPPERNONG

Vitis rotundifolia Michaux

The round to cordate, coarsely dentate leaves of this high-climbing woody vine are 6–10 cm across and turn rich yellow in the fall. The clusters of small flowers later produce tasty, purplish round berries that may be up to 2 cm in diameter.

A native of the southeastern United States north to Maryland and west to Texas, Muscadine is common in deciduous woodlands throughout the state and on our coastal dunes, where it is often a groundcover. The Cherokee used Scuppernong, as we all do today, for food.

MAY–JUNE (120-2-1)
V-5A/SCD/URG
[◗/N]

Malvaceae

SALT MARSH MALLOW

Kosteletskya virginica (L.) Presl. Ex Gray

These perennial herbs have several tall stems, 1 m or more in height, from a root crown. The alternate, triangular-ovate leaves are 6–14 cm long, and both leaves and stems are densely stellate-pubescent. The flowers are 5–8 cm across and vary from pale to deep pink; each lasts for only one day.

Native to brackish marshes of the outer coastal plain from New York to Texas, this attractive perennial is easily grown from seed and is generally hardy in the garden from the coast to the lower mountains.

JULY–OCTOBER (122-7-1)
H-5A/SNS/SRR
[○/N/$]

Malvaceae

SWAMP MALLOW; WILD COTTON

Hibiscus moscheutos Linnaeus

The conspicuous flowers of these shrubby perennial herbs, up to 2 m tall, have petals 6.5–10 cm long. The numerous anthers are united into a column around the style, a characteristic of the Mallow family. The ovate to elliptic to lanceolate leaves of Swamp Mallow are 8–22 cm long.

A native of the eastern and central United States from New York to Texas, Swamp Mallow occurs in the coastal plain and piedmont of North Carolina, where the plants grow along the edges of marshes and swamp forests. There are several horticultural varieties and hybrids of this easy-to-grow perennial.

JUNE–SEPTEMBER (122-8-2)
H-5A/SNS/URW
[◯/M/$/✪]

Malvaceae

FLOWER-OF-AN-HOUR

Hibiscus trionum Linnaeus

As the common name implies, the attractive flower of this plant, 3–6 cm broad, lasts but a short time. These branching, hairy, annual herbs are 3–5 dm tall, and the deeply palmately lobed leaves are 3–6 cm long. The calyx becomes inflated to form a paper "pod" around the maturing fruit.

These introduced Mallows may be found sporadically over much of the United States and occasionally grow in cultivated fields and along sandy roadsides, chiefly in our coastal plain and piedmont.

JUNE–OCTOBER (122-8-5)
H-5A/SOC/SRY
[◯/N/$]

Theaceae

SILKY CAMELLIA; STEWARTIA

Stewartia malacodendron Linnaeus

This shrub or small tree in the Tea Family may grow up to 6 m tall, has camellialike flowers with petals 2–4.5 cm long, and is one of North Carolina's rarest and most beautiful native shrubs. The slightly serrulate, elliptic leaves are 5–10 cm long.

These primarily southeastern plants, found from Virginia to Texas, grow in low woodlands and along creek banks, chiefly in our coastal plain but also at a few scattered localities in the piedmont and mountains. Silky Camellia is highly recommended as a landscape shrub.

MAY–JUNE (124-1-1)
S-5A/SES/SRW
[◗/M/$]

Theaceae

LOBLOLLY BAY

Gordonia lasianthus (L.) Ellis

The thick, glossy evergreen leaves of this shrub or small tree, 10–15 m tall, are often ragged where chewed by insects. The camellialike flowers are 5–7 cm across; the round, pubescent flower buds are about 1 cm in diameter.

These plants grow in the bay forests and in thickets along low ditches in the coastal plain of North Carolina, where they reach their northern limit, ranging southward to Mississippi.

JULY–SEPTEMBER (124-2-1)
S-5A/SES/SRW
[◗/M/$]

Hypericaceae

ST. ANDREW'S CROSS

Hypericum hypericoides (L.) Crantz

This shrub, 3–10 dm tall, with elliptic to linear, 8–26 mm long leaves, has distinctive flowers with four narrow yellow petals that form a cross and 2 large, leaflike sepals (the other two are tiny or absent). The 20 or more yellow-flowered *Hypericum* species growing in North Carolina can be recognized as a group by their similar flower structure—4 to 5 yellow petals and numerous stamens—and their leaves that are dotted with minute glands.

Found in dry woods, chiefly of our coastal plain and piedmont, St. Andrew's Cross ranges throughout much of the eastern United States. Native Americans used both the roots and leaves to treat a long list of ailments, including snakebite, fevers, and pain.

MAY–AUGUST (126-1-2)
S-40/SEE/SRY
[◗/D/☼]

Hypericaceae

ST. JOHN'S WORT

Hypericum prolificum Linnaeus

These coarse shrubs are up to 2 m tall, with elliptic to linear, revolute leaves, 3–5 cm long. The numerous, slender stamens give a light feathery appearance to the clustered, showy, 2.5 cm broad flowers.

Ranging through much of the eastern and central United States, this St. John's Wort grows in meadows and seepage slopes at scattered localities in our mountains and piedmont.

JUNE–AUGUST (126-1-8)
S-50/SEE/URY
[◯/M]

Cistaceae

MOUNTAIN GOLDEN HEATHER; HUDSONIA

Hudsonia montana Nuttall

This very rare small shrub, up to 15 cm tall, has needlelike leaves and solitary flowers, about 2.5 cm wide, at the ends of short branches. The linear leaves are spreading and thus differ from the closely appressed leaves of the coastal *H. tomentosa*.

Hudsonia is endemic to North Carolina and known from only a few shrub balds at high elevations in two mountain counties. It is both state and federally listed. Care should be made to protect these attractive, rare shrubs where they grow. ✗

JUNE–JULY (129-1-2)
S-5A/SLE/SRY
[❍/D/E]

Violaceae

BIRDFOOT VIOLET

Viola pedata Linnaeus

Probably the most distinctive of the Violets occurring in North Carolina, this 5–8 cm tall rhizomatous perennial has 2–5 cm broad flowers that are usually lavender, rarely white, or occasionally bicolored, in which the 2 upper petals are deep purple and the 3 lower ones lavender.

A native of the eastern and central United States, this attractive Violet grows in rocky, open woodlands and on the poor clay soil of roadsides throughout the mountains and piedmont, as well as in many of our coastal plain counties. The Cherokee made use of Birdfoot Violet as a medicine and an insecticide. Propagate this plant by seed for well-drained, sunny spots in the wild garden.

MARCH–MAY (130-2-1)
H-5B/SOL/SZB
[◗/D/$/✪]

Violaceae

PURPLE VIOLET

Viola sororia Willd.
[*Viola papilionacea* Pursh]

The smooth, heart-shaped leaves, 4–10 cm long on 2–8 cm petioles, and the fragrant, zygomorphic flowers both arise directly from the rhizome of these perennial, "stemless" violets. The purple chasmogamous flowers with a lighter eye may also be white with a bluish or green eye. Green, almost sessile, cleistogamous, self-pollinating flowers may also occur at the base of the same flowering stem.

Found throughout our state, often in large clumps or colonies in old fields and alluvial woods and along roadsides, these violets are widespread throughout the eastern United States. Although a good choice for the wild garden, Purple Violet is too aggressive for some gardeners. The Cherokee used the leaves of Purple Violet to make a spring tonic and to treat colds.

FEBRUARY–MAY (130-2-3)
H-5B/SCS/SZB
[◗/N/$/✪]

Violaceae

HAIRY VIOLET

Viola villosa Walter

As the species name implies, these small, rhizomatous perennials have densely pubescent leaves, 2.5–7.5 cm long. The flowering stems are shorter than the leaves, and, as with most other Violets, bear two kinds of flowers: open, cross-pollinated flowers with colorful petals, and closed, self-pollinated, basal flowers that have no showy petals.

These rare native plants of the southeastern United States grow in moist, sandy or rocky soil at the margins of pocosins and woodlands in a few scattered localities in all of our state's provinces. ✗

FEBRUARY–APRIL (130-2-6)
H-5B/SOS/SZB
[❍/M/R]

Violaceae

LANCE-LEAVED VIOLET

Viola lanceolata Linnaeus

These acaulescent, stoloniferous perennials have erect or suberect, toothed, lanceolate to linear leaves, 3–6 cm long. The flowers are 1–2 cm long, and the lower 3 petals with brown-purple lines near the base are beardless.

This native of eastern and central North America is found chiefly on our coastal plain. The quite similar, and closely related, *V. primulifolia* is found in similar moist habitats throughout the state.

MARCH–MAY (130-2-16)
H-5B/SNS/SZW
[○/M/$]

Violaceae

HALBERD-LEAVED YELLOW VIOLET

Viola hastata Michaux

Each year, these low, rhizomatous perennials with a leafy stem, 0.5–2.5 dm long, produce only 2 or 3 5–10 cm long, hastate leaves that usually have distinct variegated silver-gray zones between the veins.

These widespread natives of the eastern United States grow in usually moist, deciduous woods of our mountains and piedmont. This is a great choice for the woodland garden!

MARCH–MAY (130-2-19)
H-5A/SCS/SZY
[◗/M/$]

Violaceae

CANADA VIOLET

Viola canadensis Linnaeus

The 1 to 8 leafy stems of this rhizomatous perennial are 1.5–4 dm tall and have ovate to cordate leaves, 3–12 cm long. The white petals are purple on the back, and the lateral and spur petals have distinctive purple veins. The flowers turn pink after pollination. Unlike most Violets, these have no cleistogamous flowers.

Canada Violet extends from Canada throughout the northern states and south along both the east and west mountain ranges. These attractive violets often form conspicuous clumps, or colonies, in the wooded coves of our mountains and are great for the moist, shaded wild garden.

APRIL–JULY (130-2-21)
H-5A/SOS/SZW
[●/M/$/✪]

Violaceae

LONG-SPURRED VIOLET

Viola rostrata Pursh

The 1.5–2.5 cm broad flower of this interesting rhizomatous, perennial Violet is distinguished by its slender spur, 1–2 cm long. The ovate leaves are 2–4 cm long.

Natives of the eastern United States, these plants are infrequent in our area, where they grow in moist woods and hemlock forests in a few mountain counties. ✗

APRIL–MAY (130-2-23)
H-5B/SCS/SZB
[●/M/*R*]

Violaceae

FIELD PANSY

Viola bicolor Pursh
[*Viola rafinesquii* Greene]

Of the more than 20 Violet species occurring in North Carolina, this is one of the few annuals. Multiple-branched stems, 0.5–4 dm long, bear ovate leaves with prominent ciliate stipules. The flowers are 1–2 cm broad.

Rather widespread in pastures and lawns and along roadsides in the eastern and midwestern United States, this violet is frequent in our mountains and piedmont. Start a patch in your lawn—they don't hurt anything!

MARCH–MAY (130-2-28)
H-5A/SOE/SZB
[◯/N]

Passifloraceae

PASSION FLOWER; MAYPOPS

Passiflora incarnata Linnaeus

The intricate flowers of this herbaceous, trailing vine, which may be up to 8 m long, are 4–6 cm wide. The deeply 3-lobed leaves are 6–15 cm long. The small, melon-shaped, edible fruits are 4–7 cm long, and "pop" if mashed.

These native perennials of the southeastern and central United States north to Pennsylvania grow in fields and along roadsides and fencerows in every county of North Carolina. The Cherokee used the root of Passion Flower for medicine and the fruit for food. If you have a sunny area and a low fence, plant some Maypops.

MAY–JULY (131-1-1)
V-5A/SOL/SRB
[◯/N/$/✪]

Cactaceae

CACTUS; PRICKLEY PEAR

Opuntia humifusa (Raf.) Rafinesque
[*Opuntia compressa* (Salisb.) Macbride]

The flattened, fleshy green stems of this 1–4 dm tall perennial are jointed, and each joint, or "pad," is 2–8 cm broad and bears small clusters of numerous fine spines. The flowers, 5–7 cm in diameter, produce reddish, pulpy, obovoid fruits, 2–3 cm long.

These plants are broadly distributed through the eastern and central United States and grow in dry, sandy or rocky habitats at scattered localities throughout North Carolina. The severed pads root easily for dry areas in the wild garden. The pads, fruit, and seeds are edible and were widely used by Native Americans as a snakebite remedy, for food, and to fix the colors for dyeing. Warning: the spines cause severe skin irritation.

MAY–JUNE (132-1-1)
H-5A/—/SRY
[❍/D/$/✪/☞]

Melastomataceae

YELLOW MEADOW BEAUTY

Rhexia lutea Walter

These usually branched, glandular, hairy, 2–6 dm tall perennials have elliptic to oblanceolate leaves, up to 2.8 cm long and 6 mm wide. The petals are 7–12 mm long.

At its northern limit in savannas and low pinelands of our southern coastal plain, Yellow Meadow Beauty ranges along the coast to Texas. These attractive plants are highly recommended for wet, sunny areas of the wild garden.

APRIL–JULY (136-1-3)
H-40/SEE/SRY
[❍/W]

Melastomataceae

MEADOW BEAUTY

Rhexia virginica Linnaeus

The winged, pubescent stems of these branching perennials are 2–9 dm tall. The petals are 1–2.5 cm long, but the flowers last for only a few hours. The sessile, elliptic leaves are 3–7 cm long.

This widely distributed Meadow Beauty ranges throughout eastern North America and occurs in bogs, ditches, and low meadows, chiefly in our mountains and inner coastal plain. Collect some seed and sprinkle it in a moist spot in your sunny garden.

MAY–SEPTEMBER (136-1-7)
H-40/SES/SRR
[◐/w/$/✪]

Onagraceae

PRIMROSE WILLOW

Ludwigia leptocarpa (Huttall) Hara

These erect, often branched, pubescent annual or perennial herbs can be up to 1 m tall; the solitary, showy flower, 1–2 cm in diameter, has lanceolate sepals. The cylindric capsules are 2–5 cm long and differ from those of other *Ludwigia* species, which have distinctive obconic capsules.

A plant primarily of the southern states north to Pennsylvania, this species grows in marshes and ditches of our coastal plain. The 22 other species of *Ludwigia* found in North Carolina are also primarily in the coastal plain.

JUNE–SEPTEMBER (137-1-7)
H-4A/SEE/SRY
[◐/w]

Onagraceae

EVENING PRIMROSE

Oenothera biennis Linnaeus

This rather coarse, variable biennial, 0.5–2 m tall, has lanceolate leaves, 1–2 dm long, and terminal clusters of slender, conical buds opening into showy yellow flowers, 2.5–5 cm across.

Native to much of the United States and southern Canada, these somewhat weedy plants are frequent or common in old fields and along roadsides throughout the state. This common Evening Primrose has a long history of use by Native Americans for both medicine and food, and there is current interest in, and commercial production of, Evening Primrose seed oil, which is useful in the treatment of asthma, rheumatoid arthritis, and metabolic disorders.

JUNE–OCTOBER (137-2-1)
H-4A/SNE/SRY
[◐/N/$/✪]

Onagraceae

PINKLADIES

Oenothera speciosa Nuttall

This attractive, usually branched perennial, up to 7.5 dm tall, has oblanceolate to elliptic leaves that are 7.5 cm long and 3 cm wide. The large pink to white flowers are 5–7 cm broad early in the season and get smaller as blooming progresses.

Widespread in fields and waste places and along roadsides in our state, Pinkladies is found throughout the southern United States north to Pennsylvania. Tolerant of drought, this *Oenothera* does well in the garden and is often cultivated.

MAY–AUGUST (137-2-7)
H-4A/SEE/SRR
[◐/D/$]

Onagraceae

SUNDROPS

Oenothera fruticosa L.

These erect, pubescent perennials, up to 1 m tall, bear flowers 2–4 cm broad in dense terminal clusters from spring to midsummer. The alternate, elliptic to lanceolate, entire to serrate leaves are up to 11 cm long and 2.5 cm wide.

Native to dry woods, roadsides, and meadows scattered throughout our state and ranging broadly through the eastern United States, Sundrops is great for the sunny wild garden.

APRIL–AUGUST (137-2-9)
H-4A/SEE/RRY
[❍/D/$]

Onagraceae

FIREWEED

Epilobium angustifolium Linnaeus

The slender stems of these native perennials, up to 1.5 m tall, have numerous, alternate, lanceolate leaves, up to 1.5 dm long, below the elongate racemose inflorescence. The pink to rose flowers are 2–4 cm across.

A circumboreal species found throughout the northern United States, Fireweed follows the mountains south, reaching its southern limit in a few of our mountain counties in recently cleared or burned areas. Fireweed has a long record of use by Native Americans throughout its range for food, fiber, and medicine. Fireweed is also an attractive possibility for the native garden.

JULY–SEPTEMBER (137-3-2)
H-4A/SNE/RRR
[❍/N/$/✪]

Araliaceae

DWARF GINSENG

Panax trifolius Linnaeus

A close relative of Ginseng, this small native perennial, only 1–2 dm tall, has three 4–8 cm long, sessile leaflets per leaf and fruits that are more yellow than red; the roots are globose rather than elongate and are without commercial value.

These plants, which range throughout northeastern North America, grow in rich woods in only six counties in North Carolina, near the southern limit for the species. The Cherokee used Dwarf Ginseng for the treatment of many ailments, including rheumatism, gout, headache, and tuberculosis. This small plant may now have some value as a plant for the shade garden, although it is somewhat ephemeral. ✗

APRIL–JUNE (139-2-1)
H-5W/PNS/URW
[●/M/$/✪/*R*]

Araliaceae

GINSENG; SANG

Panax quinquefolius Linnaeus

These 2–6 dm tall perennials have a terminal whorl of 3 palmately compound leaves, each with 5 to 7 leaflets at the top of the stem. The long-stalked, 1–12 cm terminal umbel of small, greenish flowers is followed by bright red fruits, about 10 mm in diameter.

Once widespread throughout eastern North America, and in the rich woods of our mountains and piedmont, these overcollected plants are now quite rare throughout their range—and are state listed in North Carolina—because the dried root is in great demand as a Chinese folk medicine as well as an herbal for domestic consumption. There is a long historical record of the medicinal use of Ginseng by Native Americans. By the nineteenth century, it was also a cash crop for some tribes and for white settlers. (The first cargo ship to sail from the port of Wilmington was said to be taking 16 tons of Ginseng to China.) ✗

MAY–JUNE (139-2-2)
H-5W/MES/URW
[●/M/$/✪/*sc*]

Araliaceae

WILD SARSAPARILLA

Aralia nudicaulis L.

This rhizomatous, acaulescent, 2–6 dm tall perennial, closely related to Ginseng, has a single large, ternately compound leaf with 3 to 5 ovate to elliptic divisions up to 18 cm long and 7 cm wide. The 1 or 2 flower stalks (scapes) are 0.8–2 dm long and each bear a terminal cluster of small, umbellate flowers, and later black drupes.

Found in upland woods of our mountains, Wild Sarsaparilla occurs from the southern Appalachians through the northern United States and into Canada. Native Americans used it as both medicine and food, and European settlers made wine from its fruit and included it in various forms in nineteenth-century medicines. This attractive herb is recommended for the shade garden.

MAY–JULY (139-3-2)
H-5B/POS/URW
[●/N/$/✪]

Araliaceae

SPIKENARD

Aralia racemosa Linnaeus

This thornless, herbaceous native perennial, up to 2.5 m tall, has a large, aromatic rhizome. The clusters of small greenish-white flowers are borne in long, compound racemes. The 3 to 7 ovate leaflets may be up to 18 cm long.

Spikenard is infrequent to rare in the rich woods of our mountain counties, where it is near the southern limit of its eastern and central North American range. Used historically for medicine by Native Americans, it is also a component of modern herbal homeopathic medicine.

JUNE–AUGUST (139-3-3)
H-5A/POS/PRW
[●/M/$/✪]

Apiaceae

MARSH PENNYWORT

Hydrocotyle umbellata Linnaeus

This plant's round, peltate, succulent, aromatic dark green leaves and short, 5–10 cm flower stalks arise from elongate rhizomes. The small flowers are in a compact umbel that is usually 1–2 cm in diameter.

Native to much of eastern North America, the Pacific Coast, and tropical America, these mat-forming perennials commonly grow in moist, open areas and in roadside ditches of our coastal plain. A similar but more robust species, *H. bonariensis*, may be found at the high tide line on the beaches of a few coastal counties.

APRIL–SEPTEMBER (140-1-1)
H-5B/SRS/URW
[○/W/✪]

Apiaceae

RATTLESNAKE MASTER; BUTTON ERYNGO

Eryngium yuccifolium Michaux

Solitary or in clumps, this coarse perennial, 2.5–12 dm tall, has leathery, linear, weakly spined leaves, 1.5–8 dm long, that clasp the stem and do indeed, as their species name would have it, look a bit like Yucca leaves. The globose heads of small flowers are arranged in cymes.

Scattered throughout our state along sandy roadsides and open woods, this species is found throughout the Southeast and in the midwestern states, where it is a part of wet or dry prairies. Native Americans and early settlers used its root to make a medicine for rattlesnake bite, giving rise to the common name Rattlesnake Master. Button Eryngo makes a good addition to the wild garden for its interesting texture and form.

JUNE–AUGUST (140-4-3)
H-5B/SLE/KRW
[○/N/$/✪]

Apiaceae

BLUEFLOWER ERYNGO

Eryngium integrifolium Walter

The usually solitary, often branched, 2–8 dm stems of Blueflower Eryngo bear attractive, compact, cymose flower heads. The lanceolate to ovate, usually toothed, basal leaves are 2–10 cm long.

Inhabiting wet pinelands, meadows, and savannas, primarily of the coastal plain, and occurring throughout the southeastern states west to Texas, these interesting perennials should be considered for wetter areas of the wild garden.

AUGUST–OCTOBER (140-4-5)
H-5B/SNS/KRB
[❍/w]

Apiaceae

WILD CARROT; QUEEN ANNE'S LACE

Daucus carota Linnaeus

This common, introduced, cosmopolitan, biennial weed from central Asia is the living ancestor of our garden carrot. Plants are 4–10 dm or more tall with ovate leaves that are highly pinnatified. The flowers are arranged in a compound umbel typical of the family, with rays 1–6 cm long. The central flower of each umbel is often maroon.

Queen Anne's Lace is found in fallow fields and waste places and along roadsides throughout North Carolina and much of North America. The roots and seeds have been used medicinally for diuretic and contraceptive functions. This plant is very easy to grow in the open, wild garden.

MAY–SEPTEMBER (140-5-1)
H-5A/BLS/URW
[❍/N/$/✪/☞]

Apiaceae

ANISE ROOT; SWEET CICELY

Osmorhiza longistylis (Torr.) DC

These 4–8 dm tall native perennials have ternately-pinnately divided leaves, 1–3 dm long, and roots with a strong anise scent. The flowers are arranged in compact, compound umbels, and the slender, fusiform black fruits are barbed, attaching to fur (and clothing) for dispersal.

Sweet Cicely inhabits moist woodlands of our piedmont and mountains, extending north to Canada. A very similar and closely related species, *O. claytonii*, differing primarily by its shorter style length, is more likely to be found in our mountains. The roots of both species have been used in Native American and folk medicine as a tonic for stomachache. The sweet roots are not poisonous—but be sure of proper identification before you taste! This plant offers an attractive umbel recommended for the wild garden.

APRIL–MAY (140-8-2)
H-5A/BOS/URW
[●/M/✪]

Apiaceae

GOLDEN ALEXANDER

Zizia aurea (L.) W. D. J. Koch

These compact, brightly colored perennials, up to 8 dm in height, have twice, usually ternate, compound basal leaves. The flowering umbels have 10 to 18 rays and produce both perfect and male flowers. As with all members of the genera *Zizia* and *Thaspium*, the flowers are protogynous—the stigmas are receptive before the pollen is shed.

These plants grow in alluvial woods, swamp forests, creek bottoms, and wet meadows in our mountains and piedmont, and generally throughout the eastern and midwestern United States and Canada. The shorter-statured *Z. trifoliata*, with more open umbels, is found commonly in open woodlands throughout our piedmont and mountains. Golden Alexander is appropriate for the wild garden and, if interplanted with the semiparasitic Castilleja, can be striking. Although the roots were used by Native Americans to treat fever and, historically, were used to heal wounds, they can also be toxic.

APRIL–MAY (140-14-2)
H-5A/SOL/URY
[◗/M/$/✪/△]

Apiaceae

GOLDEN ALEXANDER; MEADOW PARSNIP

Thaspium barbinode (Michx.) Nuttall

These tall, 5–10 dm perennials with thin leaves, 2 or 3 times ternately divided, have 4 to 8 or more compact umbels of both perfect and male flowers. Fruits of *Thaspium* species are winged and thus differ from the smooth seeds of *Zizia* species.

Found in rich woods and along stream banks, chiefly in our mountains and piedmont, this species would make an attractive addition to the shade garden.

APRIL–MAY (140-15-2)
H-5A/POS/URY
[●/M/✪]

Apiaceae

POISON HEMLOCK; POISON PARSLEY

Conium maculatum L.

The smooth, hollow, branching stems of this 1–2 m tall biennial from a thick taproot are streaked with red and have an unpleasant "mousy" odor when bruised. This is a very poisonous relative of the common Queen Anne's Lace. Learn how to recognize it!

These rank European weeds, famous historically as the plant providing the poison that Socrates drank, are found spreading along roadsides and in pastures and waste places throughout our state and eastern North America. All parts of *Conium* are extremely poisonous. Do not ingest!

MAY–JUNE (140-24-1)
H-5A/POL/URW
[○/N/✪/◉]

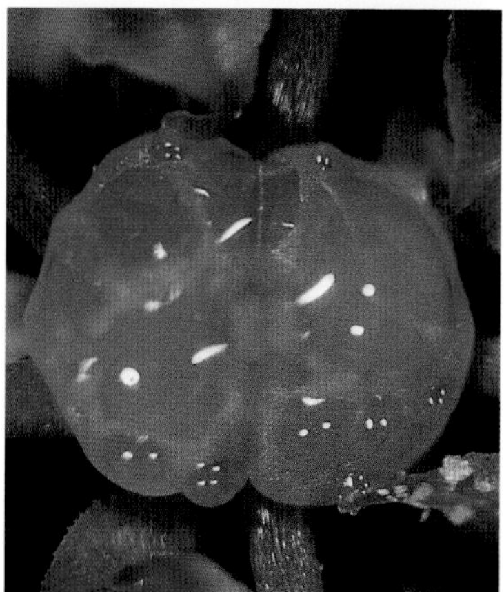

Apiaceae

FILMY ANGELICA

Angelica triquinata Michaux

This glabrous, often purple-stemmed biennial or perennial is 6–20 dm tall and has large, 3-parted, basal and lower cauline leaves with often 5-parted, coarsely serrate leaflets. The large terminal compound umbels have 15 to 25 or more rays; the greenish flowers produce copious amounts of nectar that, from many field observations, appears temporarily to drug wasp and bee pollinators.

Filmy Angelica is found in scattered localities in moist deciduous forests, along shaded roadsides, and in balds in the high mountains of the Appalachians from Georgia to Pennsylvania.

AUGUST–SEPTEMBER (140-32-2)
H-5A/POS/URG
[◗/M]

Apiaceae

COW PARSNIP

Heracleum lanatum Michaux
[*Heracleum maximum* Bartram]

These large, rank wooly, strong-scented perennials or biennials grow to 2.5 m tall and have large compound umbels, 1.5–3 dm in diameter, with 12 to 25 rays. The large pinnately or ternately compound leaves have coarsely serrated leaflets, and the petiole sheaths are conspicuously expanded.

Cow Parsnip is found at its southern limits in our mountains along stream banks, meadows, and roadsides and is more common westward and northward throughout the northern United States and Canada. Widely used as a medicine for rheumatism, headaches, and colds by Native Americans, it is now being considered for use in the treatment of psoriasis, leukemia, and AIDS. The foliage is, however, poisonous to livestock.

MAY–JULY (140-36-1)
H-5A/POS/URW
[❍/N/✪]

Cornaceae

FLOWERING DOGWOOD

Cornus florida Linnaeus

The state flower of North Carolina and the state tree of several other southeastern states, these small native trees, with opposite, entire, elliptic leaves, are 5–15 m tall, bear clusters of small, inconspicuous greenish-yellow flowers surrounded by 4 subtending white bracts, 2–5 cm long, that function as showy petals. Clusters of brilliant red "berries" (drupes) form in the fall.

Dogwood is frequent to common in open woodlands throughout the state and over much of the eastern United States and has a long history of horticultural and landscape use. The wood is very hard and has been used to make wedges to split other woods. A Dogwood root bark tea was used throughout the nineteenth century as a treatment for malaria, and Dogwood twigs were used as chewing sticks for cleaning teeth.

MARCH–APRIL (142-1-1)
T-50/SOE/URW
[◗/N/$/✪]

Cornaceae

BUSH DOGWOOD

Cornus amomum Miller

This native, opposite-leaved shrub is 1–3 m tall. The flat inflorescence of small white flowers is 5–8 cm broad and is followed in the fall by small blue, or bluish-white, fruits.

This northeastern species grows along swamp borders and in marshes and alluvial woods, chiefly in our mountains and piedmont. There is a record of medicinal use by the Iroquois and Menominee and of the use of the dried inner bark for smoking tobacco by several tribes throughout the range of this species. Bush Dogwood is a good candidate for use as a landscape shrub and in the wild garden.

MAY–JUNE (142-1-3)
S-50/SOE/URW
[◗/M/$/✪]

Clethraceae

**WHITE ALDER;
SWEET PEPPER BUSH**

Clethra alnifolia Linnaeus

These native shrubs, to 2.5 m tall, often form clumps from root suckers and have alternate, simple, finely serrate leaves that are 4–11 cm long. The bark is gray and peels with age. The very fragrant flowers are in simple or compound racemes. Our taller mountain species, *Clethra acuminata*, has leaves 8–20 cm long and flowers in simple, slender racemes.

This species is native to coastal regions from Maine to Texas and occurs in North Carolina's rich pocosins, bays, and pine barrens. Plants of both *Clethra* species are becoming more popular for garden and landscape use.

JUNE–JULY (143-1-1)
T-5A/SE
[❍/M/$]

Ericaceae

SPOTTED WINTERGREEN; PIPSISSEWA

Chimaphila maculata (L.) Pursh

The distinctive white-veined leaves of this glabrous, rhizomatous perennial subshrub, only 1–2 dm tall, are easily recognized even without the terminal inflorescence of usually 1 to 3 nodding, waxy, white flowers, 1–2 cm broad.

Spotted Wintergreen is found frequently in upland pine and deciduous woods throughout our state and much of eastern North America. The Indian name, "Pipsissewa," translated "to make water," attests to the plant's diuretic quality and, indeed, there is a long history of medicinal use of this plant both by Native Americans and European settlers.

MAY–JUNE (145-1-1)
S-5A/SNS/URW
[●/N/$/✪]

Ericaceae

INDIAN PIPES

Monotropa uniflora Linnaeus

These 1–2 dm tall pale, fleshy, saprophytic perennials, sometimes mistaken for fungi, are in the same family as *Rhododendron* and can neither be transplanted nor grown from seed. The nodding flowers, about 1.5–3 cm long, are basically similar to those of other members of the Heath family but are more compact. As the fruit matures, the stem straightens and the capsule is erect. The tiny, clasping white leaves, 5–15 mm long, are inconspicuous.

This widespread and easily recognized wild flower is found in deciduous woods at scattered localities throughout our state and much of North America. Indian Pipes were used for both Native American and folk medicine. ✗

JUNE–OCTOBER (145-3-1)
H-5N/SEE/SFW
[●/M/✪/*S*]

Ericaceae

PINESAP

Monotropa hypopithys Linnaeus

These colorful perennial saprophytes, 1–3 dm tall, like Indian Pipes, grow on decaying wood buried in the soil or leaf mold and cannot be grown from seed or transplanted. The smaller and more numerous flowers and the tawny yellow to red coloration make separation of the two species easy. The tiny, clasping leaves, only 5–13 mm long, lack chlorophyll.

Pinesap grows sporadically (probably depending on nutrient availability) in deciduous woodlands at scattered localities more or less throughout North Carolina and much of North America. ✗

MAY–OCTOBER (145-3-2)
H-5N/SEE/RFR
[●/M/✪/s]

Ericaceae

ROSEBAY; GREAT LAUREL

Rhododendron maximum Linnaeus

This large shrub, or rarely a small tree, up to 10 m tall, has leathery, evergreen (up to 7 years!) leaves, 1–3 dm long, and large clusters of pale pink flowers, 2–3 cm in diameter, with a short tube. These shrubs alternate a year of bloom with a year of growth.

Primarily of Appalachian distribution from Nova Scotia to Georgia, these plants grow along streams in cool forests in our lower mountains and the piedmont and, in the right habitat, are sometimes used in landscaping. Native Americans used the leaves in medicine to alleviate pain and heart trouble, but all parts of the plant are highly toxic! Do not ingest!

JUNE–JULY (145-5-1)
S-5A/SBE/UCW
[◗/M/$/✪/◉]

Ericaceae

PURPLE RHODODENDRON

Rhododendron catawbiense Michaux

This handsome, often cultivated shrub, 2–4 (but sometimes up to 6) m tall, has darker, rose-lavender flowers, shorter leaves, mostly 5–15 cm long, and a shorter stature than Rosebay.

Frequent, often in large stands, on the higher rocky slopes, ridges, and balds in our mountains and piedmont, these shrubs are endemic to the southern Appalachians from West Virginia to Alabama. As with all Rhododendron species, all parts of the plant are highly toxic. Do not ingest!

APRIL–JUNE (145-5-2)
S-5A/SBE/UCR
[❍/M/$/◉]

Ericaceae

CAROLINA RHODODENDRON

Rhododendron minus Michaux

This small evergreen Rhododendron is 1–3 m tall, with thick, coriaceous leaves, 4–12 cm wide, that are characteristically glandular-punctate beneath, as are the surfaces of the calyx and corolla. Flowers vary from pink to white.

Found along stream banks and on wooded slopes, this attractive Rhododendron is common in our mountains but less frequent in the piedmont, ranging only to nearby states. This smaller Rhododendron is an excellent landscape plant. It is also highly toxic. Do not ingest!

APRIL–JUNE (145-5-3)
S-5A/SEE/UCR
[◗/N/$/◉]

Ericaceae

PINKSHELL AZALEA

Rhododendron vaseyi Gray

These native, deciduous shrubs, up to 3 m tall, have bright pink, funnel-form flowers, with slender, exserted stamens and a style and a short corolla tube, 2–5 mm long. The flowers open before the leaves come out on these shrubs, which are state listed as significantly rare. The alternate elliptic leaves are 3.5–12 cm long.

Native to only ten mountain counties of the high western mountains of North Carolina (but relatively easily cultivated elsewhere), these attractive shrubs are found in bogs and spruce forests at high elevations. All of the Azaleas are poisonous when ingested, as are the evergreen Rhododendrons. ✗

MAY–JUNE (145-5-4)
S-5A/SEE/UCR
[◗/N/$/◉/*SR*]

Ericaceae

FLAME AZALEA

Rhododendron calendulaceum (Michx.) Torrey

Probably the most widely cultivated of North Carolina's native Azaleas, these deciduous shrubs, up to 3–4 m tall, have brilliant flowers with glandular-pubescent corolla tubes, 1.5–2.5 cm long. Flowers appear before the leaves and vary from a pale yellow to a brilliant orange-red and add splashes of color to spring forests. The obovate leaves expand after flowering to 5–10 cm long.

A native Appalachian species ranging from New York to Alabama, these plants grow in deciduous forests and on forest margins in our mountains. The Cherokee used the peeled and boiled twigs for medicine, but note that all parts of these plants are poisonous when ingested.

MAY–JULY (145-5-5)
S-5A/SBE/UTY
[◗/N/$/✪/◉]

Ericaceae

PINXTERFLOWER

Rhododendron periclymenoides (Michx.) Shinners
[*Rhododendron nudiflorum* (L.) Torrey]

The flowers of this 1–2 m tall, native, deciduous shrub appear before the leaves expand. The fragrant flowers have hairy corolla tubes that are 1.3–2 cm long. The thin elliptic leaves are 5–10 cm long when fully expanded.

Ranging from New Hampshire to Georgia and Alabama, these plants are relatively frequent, though widely scattered, along streams in deciduous forests and in low woodlands more or less throughout our state. All native Azaleas make wonderful landscape plants, but all parts of the plants are poisonous.

MARCH–MAY (145-5-9)
S-5A/SEE/UTR
[◗/M/$/◉]

Ericaceae

DWARF AZALEA

Rhododendron atlanticum (Ashe) Rehder

These stoloniferous, weakly branched shrubs are 3–5 dm tall and often form large colonies. The fragrant white flowers, with corolla tubes 1.5–2.5 cm long, appear before the leaves, which will expand to 2.5–6 cm long.

A coastal species found from New Jersey to Georgia, these shrubs are frequent to common along woodland borders, on savannas, and in low, moist areas throughout our coastal plain. These shrubs are worth a try in the wild garden just for the fragrance.

APRIL–MAY (145-5-10)
S-5A/SBE/UTW
[◗/M/$/◉]

Ericaceae

CLAMMY HONEYSUCKLE

Rhododendron viscosum (L.) Torrey

This large, attractive, branched shrub grows up to 2 m or more tall. The leaves, well developed at flowering time, are sticky or clammy because of their glandular pubescence. The white flowers have a long, slender tube, 2–3 cm in length, and a wonderful spicy fragrance.

Distributed in coastal states from Maine to Texas, these plants grow on the borders of streams, ponds, and bogs at numerous scattered localities, essentially throughout North Carolina. These plants are beautiful for landscaping but toxic if ingested.

MAY–JULY (145-5-11)
S-5A/SEE/UTW
[◗/D/$/◉]

Ericacaee

MINNIE-BUSH

Menziesia pilosa (Michx.) Juss.

The branches of this 1–2 m tall deciduous shrub are pilose to pubescent, and often stipitate-glandular. The leaves are alternate, 3–8 cm long, and pilose on both surfaces, with a very characteristic leaf tip hardened into a short, whitish spur. The urceolate corolla is 6–10 mm long; the fruit is a capsule.

Native in bogs and on thinly wooded slopes, Minnie-Bush is found in the mountains from Georgia to Pennsylvania and is recommended for the wild garden.

MAY–JULY (145-6-1)
S-4A/SEE/UCR
[◗/N/$]

Ericaceae

SAND MYRTLE

Leiophyllum buxifolium (Berg.) Elliott

This is a somewhat dimorphic evergreen shrub, sometimes low and spreading in the mountains (1–5 dm tall) and usually rather tall and erect (up to 2 m in height) in the coastal plain, with smooth, shiny leaves 0.5–2.5 cm long. The small flowers have separate petals, 3–5 mm long, that may be tinged with pink.

Sand Myrtle is found from the New Jersey Pine Barrens to the Carolinas, Georgia, Tennessee, and Kentucky. The near-prostrate form, sometimes given varietal status, grows in dry, rocky areas in a few counties of our piedmont and mountains, and the taller form grows in sandy woods of our coastal plain.

MARCH–MAY (145-7-1)
S-5A/SEE/URW
[◗/N/$/✪]

Ericaceae

MOUNTAIN LAUREL; IVY

Kalmia latifolia Linnaeus

The alternate, leathery evergreen leaves of this shrub, 2–3 m tall, are elliptical and mostly 5–12 cm long. The corolla, of 5 fused petals, is 2–3 cm broad. The anthers are tucked into the fused corolla until the pollen is ready; the pollen is then released explosively when visiting bees forage for nectar and trip the catapult.

Found throughout eastern North America growing on balds and in deciduous woods, Mountain Laurel is frequent or common in our mountains and infrequent to rare in the piedmont and inner coastal plain. There are many horticultural varieties of these attractive, hardy, widely used shrubs. The leaves were used for pain and rheumatism treatments by Native Americans and, in minute amounts, historically, to treat fever, heart conditions, and inflammation. Note: the leaves (and possibly even honey from the nectar) of these shrubs are toxic. Do not ingest!

APRIL–JUNE (145-8-1)
S-5A/SEE/UCR
[◗/N/$/✪/◉]

Ericaceae

SHEEP LAUREL; WICKY

Kalmia carolina Small
[*Kalmia angustifolia* var. *caroliniana* (Small) Fernald]

The elliptic, 2.5–7 cm long leaves of this slender, native, rhizomatous shrub, up to 1.5 m tall, are usually opposite or whorled. The smaller flowers, 1–1.5 cm broad, and their darker rose color easily distinguish this species from Mountain Laurel.

This southeastern species occurs both in rocky woodlands and bogs in our mountains and in sandy woodlands, savannas, and bogs in the coastal plain. It is a potentially valuable horticultural plant. As is Mountain Laurel, Sheep Laurel is highly poisonous.

APRIL–JUNE (145-8-2)
S-50/SEE/UCR
[❍/N/$/◉]

Ericaceae

ZENOBIA; HONEYCUPS

Zenobia pulverulenta (Bartr.) Pollard

This very attractive shrub is up to 2 m or more in height, with alternate, tardily deciduous, elliptic leaves, 3.5–8 cm long. The fragrant flowers have a broadly campanulate corolla, 7–12 mm long.

These handsome plants, native to the coastal plain from Virginia to Georgia, grow in bogs and in low thickets of woody evergreens known as "bays" and should be considered for landscape use.

APRIL–JUNE (145-9-1)
S-5A/SEE/UCW
[◗/M/$]

Ericaceae

FETTER-BUSH; MOUNTAIN ANDROMEDA

Pieris floribunda (Pursh) Bentham & Hooker

This erect, freely branching shrub, usually about 2 dm but up to 2 m tall, has alternate evergreen leaves that are 3–7 cm long. The numerous, strongly urceolate, or urn-shaped, flowers are in a terminal panicle of racemes.

Mountain Andromeda is native only to the southeastern Appalachians from West Virginia to Georgia and is found in our area only on the high mountain balds in a few western counties. ✗

MAY–JUNE (145-10-1)
S-5A/SND/PCW
[❍/N/$/◉/R]

Ericaceae

LYONIA; FETTER-BUSH

Lyonia lucida (Lam.) K. Koch

The young branches of this evergreen shrub, up to 2 m tall, are sharply 3-angled, and the urn-shaped flowers, 7–9 mm long, are borne in short, axillary clusters. The leaves have a prominent midrib and a revolute, or slightly rolled, margin.

Native to the coastal plain of the southeastern United States from Virginia to Louisiana, Lyonia grows in low woods, pocosins, and savannas of our eastern counties. A related species, *Lyonia ligustrina*, occurs in a variety of moist habitats, essentially throughout the state. All Lyonia species are highly toxic; do not ingest!

APRIL–JUNE (145-11-3)
S-5A/SEE/UCR
[◗/M/$/◉]

Ericaceae

DOG HOBBLE; LEUCOTHOE

Leucothoe fontanesiana (Steud.) Sleumer
[*Leucothoe axillaris* (Lam.) D. Don]

The tough, clustered, arching stems of this colonial, native evergreen shrub are up to 1.5 m long and will stop the hounds (but not the bear they are chasing). The lanceolate leaves are 5–13 cm long. The axillary racemes are 2–7 cm long, with small, urceolate flowers, 5–7 mm long.

Primarily found in the southern Appalachians, Dog Hobble grows along streams and in heath thickets, primarily of the mountains and upper piedmont. The Cherokee used the leaves to treat rheumatism. These plants are highly toxic; do not ingest!

APRIL–MAY (145-12-2)
S-5A/SNS/RCW
[◗/M/$/✪/◉]

Ericaceae

FETTER-BUSH

Leucothoe recurva (Buckl.) Gray

These shrubs, up to 4 m tall, have deciduous leaves, 4–13 cm long, which add spectacular rich reds to the fall color show. The urceolate or urn-shaped corolla, 6–8 mm long, consists of 5 fused petals with only the small, recurved tips separate.

Native to the Appalachians from Ohio to Georgia, these plants grow in bogs and rocky woods in our mountains. A somewhat similar species, also with deciduous leaves, *L. racemosa*, grows chiefly in our coastal plain and piedmont.

APRIL–JUNE (145-12-3)
S-5A/SES/RCW
[◗/M/$]

Ericaceae

SOURWOOD

Oxydendrum arboreum (L.) DeCandolle

This medium-sized to large tree, up to 20 m tall, has alternate, lanceolate serrate leaves, up to 2 dm long. The numerous small, urn-shaped flowers, 5–8 mm long, that produce the nectar from which prized, water-clear Sourwood honey is derived, are borne in profusion in flat, terminal sprays. The sprays of fruits and brilliant red leaves can be spectacular in the fall.

Widespread over much of the eastern United States north to New York, and often cultivated, these native trees are relatively frequent in open, well-drained, deciduous woodlands, essentially throughout our state. Sourwood was used as medicine, for food, and for fiber by the Cherokee, and, until the middle of the last century, the trunks of young Sourwood trees (15–20 cm in diameter) were cut in the dark of the January moon and air-dried under cover for one year, then used for the runners of the big horse- , mule- , or ox-drawn land sleds used to move supplies and freight on narrow, mountain trails from river landings, rail heads, or main roads, to isolated coves and hollows in the mountains.

JUNE–JULY (145-13-1)
T-5A/SNE/PCW
[◗/N/$/✪]

Ericaceae

TRAILING ARBUTUS; MAYFLOWER

Epigaea repens Linnaeus

The branched, woody stems, 2–4 dm long, of this prostrate perennial have alternate, ovate evergreen leaves, 3–8 cm long. The small, 6–12 mm long, urceolate, very fragrant flowers are in a congested raceme or spike and bloom in early spring.

Trailing Arbutus grows in dry, sandy or rocky woodlands at scattered localities throughout our state and much of eastern North America. There are only two species of *Epigaea* worldwide, the other occurring in Japan. The leaves of this plant were used in both Native American and folk medicine as a treatment for kidney disorders. Do not dig! These plants will not transplant and are virtually impossible to grow. Enjoy them where they are. ✗

FEBRUARY–MAY (145-15-1)
S-5A/SEE/RFR
[◗/D/✪]

Ericaceae

WINTERGREEN; CHECKERBERRY

Gaultheria procumbens Linnaeus

Also called Teaberry, these low, upright, evergreen, rhizomatous perennials, 1–2 dm tall, have very aromatic leaves. The small urceolate flowers, 6–12 mm long, are followed by fleshy, red, berrylike fruits.

Wintergreen is found through much of the eastern United States south to Georgia in varied woodland habitats, chiefly in our mountains and coastal plain. Although this species is an attractive plant for the garden, it is hard to grow and is best propagated by seed. Both traditional and modern use is made of Wintergreen for medicine, food, and flavoring. Warning: the essential oil that is the wintergreen flavor is highly toxic and can be absorbed through the skin.

JUNE–AUGUST (145-16-1)
H-5A/SED/SCW
[●/D/$/✪/◉]

Ericaceae

DWARF HUCKLEBERRY

Gaylussacia dumosa (Andrz.) T. & G.

These low, rhizomatous shrubs, 1–4 dm tall, have resinous glands on the underside of their deciduous to semi-evergreen leaves. The globular, 10-seeded, black berries are 7–10 mm in diameter and are glandular-puberulent. These fruit characters separate *Gaylussacia* from *Vaccinium*.

Dwarf Huckleberry grows in the sandy habitats of the coastal plain and adjacent piedmont throughout North Carolina and the eastern United States. Three other Huckleberries are found in our state from the mountains to the coast.

MARCH–JUNE (145-18-2)
S-5A/SEE/RCW
[◗/D/$]

Ericaceae

GOOSEBERRY; SQUAW-HUCKLEBERRY; DEERBERRY

Vaccinium stamineum L.

The unusual flowers of this large shrub, 1–5 m tall, do not form buds: all of the flower parts just grow "in the open" from minute rings of meristematic floral tissue. The characteristic presence of leaflike bracts, often as large as the leaves, in the racemose inflorescence always identifies this species, as does the large green berry, 0.7–1.5 cm in diameter.

Common throughout the state in relatively dry, rocky or sandy woods and more broadly throughout the eastern and central United States, the fruits are an important food for wildlife.

APRIL–JUNE (145-19-2)
S-5A/SEE/RCW
[◗/D/$]

Ericaceae

HIGHBUSH BLUEBERRY

Vaccinium corymbosum Linnaeus

The cultivated Blueberries of commerce were developed from these variable native shrubs, 1–4 m tall, which, like the other Blueberries, have small, urn-shaped flowers, 8–11 mm long. The smooth, elliptic leaves are 4–8 cm long.

Native to much of the eastern United States, Blueberries grow in pine or deciduous woodlands and along woodland borders in the mountains and in low areas of our coastal plain. Over the past few decades the importance of blueberries as a crop plant (and as a landscape plant) has increased tremendously.

FEBRUARY–MAY (145-19-8)
S-5A/SEE/RCW
[◗/M/$]

Ericaceae

BEARBERRY

Vaccinium erythrocarpum Michaux

This upright shrub, 3–15 dm tall, can be easily identified by the distinctive, slender, red, 4-parted flowers. The deciduous, lanceolate leaves, 3–8 cm long, are finely serrate.

A native of the southern Appalachians from West Virginia to Alabama, these plants are occasionally found in our high mountains where they grow in moist woodlands and bogs. The juicy dark red berries, 7–9 mm long, are seldom seen but are part of the woodland mast crop that supports forest wildlife.

MAY–JULY (145-19-13)
S-4A/SNS/SCR
[●/M/$]

Ericaceae

CRANBERRY

Vaccinium macrocarpon Aiton

This trailing shrub, usually rooting at the nodes, has erect branches, up to 2 dm tall, with small, elliptic leaves, 5–18 mm long, and is among the native members of the Heath family that have been domesticated. The small flowers with sharply recurved petals are followed in the fall by the edible red berries, 1–1.5 cm in diameter.

Cranberry ranges through northeastern and central North America south to North Carolina, where it is found in bogs and pocosins of our mountains and coastal plain. Since some of these small bogs have been lost to development and highway construction in the last decade or so, Cranberry is now listed as significantly rare in North Carolina and is known from only nine counties. The Montagnais of Quebec used branches of Cranberry as a medicine for pleurisy, but most records of Native American use is as a food. The berries, rich in vitamin C, were often carried aboard sailing ships (packed in barrels of fresh water) and eaten to prevent scurvy. ✗

MAY–JULY (145-19-14)
S-4A/SEE/SCR
[❍/W/$/✪/*SR*]

Diapensiaceae

OCONEE BELLS

Shortia galacifolia Torrey & Gray

As the specific name implies, this low, attractive, rhizomatous perennial has glossy evergreen leaves, 3–8 cm long, resembling those of *Galax*, which is also a member of the same small family. The bell-shaped, 5-petaled white flowers are 2–2.5 cm long on scapes up to 18 cm long.

This rare, state-listed plant grows along stream banks in rich woods in a very few counties in the mountains of North and South Carolina and Georgia. Fortunately, Shortia is relatively easy to grow in the proper woodland habitat, and nursery propagated plants may save it from extinction. Buy a few plants to get a start of it for the shady wild garden. ✗

MARCH–APRIL (146-1-1)
H-5B/SRS/SCW
[●/M/$/*E-SC*]

Diapensiaceae

PIXIE MOSS

Pyxidanthera barbulata Michaux

This inconspicuous, 1–3 cm high, prostrate, creeping, woody perennial with evergreen, lanceolate leaves, 4–8 mm long, has small, sessile flowers with lobes, 3–5 mm long.

Pixie Moss is rare and irregularly distributed in the dry sandy woods and pine barrens of the coastal plain from northern South Carolina to New Jersey. In North Carolina, specimens have been found chiefly south of the Neuse River. A rarer variety with smaller, more pubescent, leaves, the Sandhills Pixie Moss (*P. brevifolia*), is known from only four counties in the sandhills and is listed as endangered in the state. ✗

MARCH–APRIL (146-2-1)
H-5A/SBE/SRW
[○/D/*R*]

Diapensiaceae

GALAX

Galax urceolata (Poir.) Brummitt
[*Galax aphylla* Linnaeus]

The leathery, round to ovate leaves, 3–10 cm long, of these low, rhizomatous perennials turn reddish-brown or bronze in the winter. The flowers are in a 5–10 cm long terminal raceme on a 2–4 dm tall flower stalk.

Native to the southern Appalachian region but cultivated elsewhere, Galax grows in open, often rocky deciduous woods in our mountains and, less frequently, in the piedmont and coastal plain. The Cherokee used an infusion of the root to treat kidney ailments. Often used in Christmas decorations, these plants have been threatened by overcollection in many localities. In some communities, local governments have prohibited the sale of Galax leaves. They do not transplant easily, so please do not dig! ✗

MAY–JULY (146-3-1)
H-5B/SOS/RRW
[●/N/$/✪]

Primulaceae

SHOOTING STAR

Dodecatheon meadia Linnaeus

This often cultivated, herbaceous perennial has smooth, lanceolate leaves, 1–3 dm long, and a flowering stalk, 2–6 dm tall. The flowers are normally white but may be pale pink or lilac.

Primarily a woodland plant found in partial shade on neutral or basic soils, Shooting Star ranges through much of the eastern and central United States north to New York and Minnesota. In North Carolina it occurs at a few scattered constantly moist but well-drained localities in the mountains and piedmont.

MARCH–MAY (147-1-1)
H-5B/SNE/URW
[◗/M/$]

Primulaceae

FRINGED LOOSESTRIFE

Lysimachia ciliata Linnaeus

The opposite, ovate to lanceolate, 6–12 cm long leaves of this rhizomatous, 2–13 dm tall perennial have fringed petioles. A usually solitary, slender-stalked flower, about one inch across, is in the axil of each of the upper leaves.

Native through much of North America, Fringed Loosestrife grows in alluvial woods or on moist slopes in many widely separated areas, chiefly in our mountains and piedmont.

JUNE–AUGUST (147-2-1)
H-5O/SOE/SRY
[◗/M/$]

Primulaceae

WHORLED LOOSESTRIFE

Lysimachia quadrifolia Linnaeus

The leaves of these 3–10 cm tall, unbranched perennials are in whorls of 4 to 6. A single flower, 1.3–1.6 cm broad, is borne on a very slender stalk arising from the axil of each of the upper leaves.

A species of the northeastern United States south to Georgia and Alabama, irregularly distributed throughout North Carolina, these plants usually grow in full sun in clearings or along woodland margins. The Cherokee and Iroquois used roots of Whorled Loosestrife to treat kidney problems and as an emetic.

MAY–JULY (147-2-6)
H-5W/SEE/SRY
[○/M/$/✪]

Symplocaceae

HORSE-SUGAR; SWEETLEAF

Symplocos tinctoria (L.) L'Heritier

This deciduous shrub or small tree, up to 10 m tall, which usually flowers before the leaves appear, is sometimes cultivated. The simple, alternate leaves are 8–13 cm long. The compact clusters of small, fragrant, cream-colored flowers, borne in profusion along the branches of the previous year's growth, are quite conspicuous in the early spring woods.

Found throughout the Southeast north to Delaware, Horse-Sugar grows in sandy thickets, in alluvial woods, and along streams in all parts of North Carolina except the upper piedmont.

MARCH–MAY (151-1-1)
T-5A/SEE/KCY
[◗/M/$/✪]

Styracaceae

CAROLINA SILVERBELL

Halesia carolina L.

This beautiful small tree, up to 10 m tall, has clusters of 3 to 5 pendant, campanulate, white, 4-petaled flowers, 15–25 mm long. The distinctive brown, dry, beaked, 4-winged fruit is 1- to 3-seeded.

Silverbell is found in rich deciduous woods, river bottoms, and stream banks, chiefly in our mountains and upper piedmont, and more generally through the Southeast. Silverbell is a prize landscape tree and, with considerable patience (and luck!), can be propagated by seed. However, nursery-grown seedlings are not expensive and are well worth the time saved.

MARCH–MAY (152-1-1)
T-4A/SES/UFW
[◗/M/$]

Styracaceae

STORAX; SNOWBELL

Styrax americana Lamarck

This attractive bushy shrub, up to 3 m tall, has glossy, deciduous leaves, 3–7 cm long, and nodding, 5-petaled flowers, 10–15 mm long.

Native to the southeastern United States north to Illinois and Ohio, Storax grows in swamp forests, in low woods, and along stream banks of our coastal plain. Snowbell is commonly used in landscaping.

APRIL–JUNE (152-2-2)
S-5A/SBE/SCW
[◗/M/$]

Oleaceae

FRINGE TREE; OLD MAN'S BEARD

Chionanthus virginicus Linnaeus

This deciduous, often cultivated shrub or small tree, up to 6 m tall, with smooth, opposite, ovate to elliptic leaves, 8–15 cm long, has large, showy, drooping clusters of small, fragrant flowers with narrow, linear petals, 15–30 mm long.

Primarily limited to the coastal states of the southeastern United States north to New York, Fringe Trees grow in dry woods more or less throughout our state and, if in full sun and well branched, can be striking in bloom. Native Americans used decoctions of the root as a wash for sores, and, in the late nineteenth century, physicians used a tincture of the bark to treat a variety of conditions, including jaundice and rheumatism.

APRIL–MAY (153-3-1)
S-40/SEE/PRW
[◗/N/$/✪]

Loganiaceae

YELLOW JESSAMINE

Gelsemium sempervirens (L.) Aiton

These trailing or climbing woody vines, up to 5 m long, have slender, wiry stems and evergreen, lanceolate leaves, 4–7 cm long. The fragrant flowers, 2–3.8 cm long, often occur in sufficient numbers to make the plant quite showy.

Indigenous to the southeastern United States from Virginia to Texas, this vine is frequent in thickets, in open woodlands, and along roadsides throughout our coastal plain. Yellow Jessamine is the state flower of South Carolina and is wonderful for the wild garden, especially in combination with Coral Honeysuckle. Although Yellow Jessamine roots have been used as a cancer remedy in folk medicine, all parts of the plant are highly toxic. Do not ingest!

MARCH–APRIL (154-1-1)
V-50/SNE/SFY
[❍/M/$/✪/◉]

Gentianaceae

SABATIA; ROSE-PINK

Sabatia angularis (L.) Pursh

The 3–7 dm tall stem of this native biennial is strongly 4-angled, as indicated by the specific name. The showy, fragrant flowers have petals 1.5–2 cm long.

Widespread over much of the Southeast north to New York but infrequent in North Carolina, Rose-pink may be found occasionally along woodland borders and marshes at scattered localities, primarily in our piedmont and lower mountains. Sabatia is an ideal plant for the sunny wild garden. Collect and scatter seed where you want a patch of pink.

JULY–AUGUST (155-1-5)
H-50/SOE/PRR
[❍/M/✪]

Gentianaceae

MARSH PINK; MARSH ROSE GENTIAN

Sabatia dodecandra (L.) BSP

This 3–7 dm tall perennial, with cauline, narrowly lanceolate to elliptic leaves, 2–5 cm long, forms attractive clumps from short rhizomes. The distinctive flowers, up to 8 cm broad with 6 to 13 lobes, are in diffuse panicles.

As the common name implies, Marsh Pink is found in marshes and savannas; it occurs in our outer coastal plain and ranges along the coast from New York to Texas. These attractive plants are well worth the effort to grow in a wet spot of your sunny garden.

JUNE–AUGUST (155-1-10)
H-50/SNE/PRR
[❍/W/$]

Gentianaceae

FRINGED GENTIAN

Gentianopsis crinita (Froel.) Ma
[*Gentiana crinita* Froelich]

This annual or biennial Gentian, 1.5–9 dm tall, bears attractive flowers, 3.5–6 cm long, with 4 fringed petals on naked stalks, 4–10 cm long. The opposite, lanceolate, sessile leaves are 1–5 cm long and 3–17 mm wide.

These rare plants, on the state endangered list, are known from damp meadows and seepage slopes in only two of our mountain counties, where they are near the southern limit of their range. These plants can be propagated by seed and do well with proper care in the sunny garden. Do not pick, dig, or disturb (except to collect the seed), and leave for others to enjoy! ✗

SEPTEMBER–OCTOBER (155-2-1)
H-40/SNE/SFB
[❍/M/$/*E-SC*]

Gentianaceae

AGUEWEED; STIFF GENTIAN

Gentiana quinquefolia Linnaeus

The rather narrow, erect, 5-lobed flowers of this 1.5–8 dm annual are 14–24 mm long, and the angled stems and branches are slightly winged. The ovate to lanceolate leaves are 3–8 cm long.

A species of the eastern United States from southern Ontario south to Georgia, Agueweed is restricted in North Carolina to the mountains where it grows on moist road banks and the margins of streams or bogs. Despite its name of Agueweed, this species has not been used to treat fever, but rather digestive ailments. This is a neat little Gentian to get started, from seed, in your wild garden—or lawn. ✗

AUGUST–OCTOBER (155-2-2)
H-50/SNE/UTB
[◗/M/$/✪]

Gentianaceae

SAMPSON'S SNAKEROOT

Gentiana villosa Linnaeus

These glabrous perennials, 1.5–5 dm tall, are smooth and not villous or hairy as the specific name mistakenly implies. The wide leaves are 4–8.5 cm long.

Native through much of the Southeast north to Pennsylvania and Ohio, this rare Gentian is found at scattered localities in open woods and pinelands in all three provinces of our state. Leave these rare plants for others to enjoy. ✗

AUGUST–NOVEMBER (155-2-4)
H-50/SEE/UFY
[◗/M/$/*R*]

Gentianaceae

SOAPWORT GENTIAN

Gentiana saponaria Linnaeus

These 3–9 dm tall perennials with large terminal flowers, 3–4 cm long, do not produce a lather but derive their common name from the resemblance of their foliage to that of *Saponaria*, or Soapwort.

A rare plant native to the eastern United States from New York to Texas, this Gentian occurs in bogs, marshes, and low ditches at widely scattered localities in a few counties in each of our three provinces. ✗

SEPTEMBER–NOVEMBER (155-2-8)
H-50/SEE/UFB
[◗/M/$/*R*]

Gentianaceae

COLUMBO

Frasera caroliniensis Walter
[*Swertia caroliniensis* (Walter) Kuntze]

These rare, 1–3 m tall, glabrous, monocarpic perennials have whorled, lanceolate, 1.5–3.5 dm long leaves. The striking 4-petal cream flowers are marked with green, and each petal has a conspicuous green gland at the base.

Found in rich woodlands of only two of our mountain counties, and state listed, Columbo is thinly but broadly distributed in the eastern states, in which it is nowhere frequent, and the central states. Several other *Frasera* species are relatively more abundant in the western United States. The Cherokee used the root of Columbo to relieve indigestion and as an antiseptic. Do not dig; collect ripe seed in season and grow your own! ✗

MAY–JUNE (155-3-1)
H-4W/SNE/PRG
[◗/N/✪/*SR*]

Gentianaceae

PENNYWORT

Obolaria virginica Linnaeus

This low perennial, only 3–15 cm tall, has thick, rounded, opposite, purplish-green leaves just 5–15 mm long. The small white to lavender flowers, 7–15 mm long, are solitary or in clusters of 3 in the axils of the upper leaves.

Essentially a species of rich, deciduous forests of the southeastern and central United States north to New Jersey, Pennywort is found chiefly in our mountains and northern piedmont, where it is often overlooked because of its inconspicuous coloration. The Cherokee used Pennywort to treat colds and colic, and the Choctaw used a root decoction as a wash for cuts and bruises.

MARCH–MAY (155-5-1)
H-4O/SOE/UFW
[●/M/✪]

Apocynaceae

BLUE STAR

Amsonia tabernaemontana Walter

This slender, leafy perennial has one to several, 4–10 dm tall stems from a single root crown. The panicle of attractive blue flowers with floral tubes, 6–8 mm long, terminates the stem. The ovate to lanceolate leaves are 6–12 cm long.

Native to the eastern and central United States, *Amsonia* grows on wooded slopes and in bottomlands, primarily in our piedmont and is an ideal plant for full sun or partial shade in the wild garden.

APRIL (156-1-1)
H-5A/SOE/PTB
[◗/M/$]

Apocynaceae

SPREADING DOGBANE

Apocynum androsaemifolium Linnaeus

The large, fragrant, almost showy flowers of this native perennial, 2–5 dm tall, are 4–7 mm long and distinguish it immediately from *A. cannabinum*. The fruits of both are slender follicles 1.2–2 dm long but only about 2–3 mm in diameter.

These perennials, widely distributed throughout North America south to our area, grow along roadsides and woodland margins only in our mountains, where they add their rich yellow color to the landscape, or your native garden, in the fall. Spreading Dogbane was used as a remedy for many illnesses by Native Americans and has shown some antitumor activity, but this species is toxic. Do not ingest!

JUNE–AUGUST (156-3-3)
H-50/SOE/UCW
[◗/N/✪/◉]

Asclepiadaceae

SWAMP MILKWEED

Asclepias incarnata L.

These native perennials, which produce milky sap, may have one or several 5–15 dm stalks from a single rootstock. The attractive, intricate flowers are arranged in globose clusters 2.5–4.5 cm in diameter. The opposite, lanceolate leaves are 6–15 cm long.

Plants occurring in North Carolina are of the subspecies *pulchra*, which is distributed from New England through the eastern coastal states to Texas. It is a conspicuous but not common plant of marshes and moist meadows in our mountains, piedmont, and northern coastal plain and is an excellent plant for the wild garden. The empty pods add interest to dried arrangements. Swamp Milkweed was used by both Native Americans and colonists for medicine, but note that this species is toxic if ingested.

JULY–SEPTEMBER (157-1-1)
H-50/SNE/URR
[❍/M/$/✪/⚠]

Asclepiadaceae

BUTTERFLY WEED; PLEURISY-ROOT

Asclepias tuberosa Linnaeus

Unlike our other species of *Asclepias*, this perennial has clear sap and alternate, 4–10 cm long leaves. Often cultivated for its brilliant flowers, it may have several 2–8 dm stems from a single root crown.

Native over much of the eastern United States, Butterfly Weed is relatively frequent in dry fields, on roadsides, and along woodland margins throughout our area. This colorful milkweed makes an excellent addition to a wild flower garden and will attract both bees and butterflies to its bright flowers. Widely used medicinally for lung-related problems among other ailments (hence the name Pleurisy-Root), these plants are toxic when ingested. Butterfly Weed is easily grown from seed for your wild garden—or just for the color and butterflies.

MAY–AUGUST (157-1-4)
H-5A/SEE/URY
[○/D/$/✪/⚠]

Asclepiadaceae

COMMON MILKWEED

Asclepias syriaca Linnaeus

These coarse, rhizomatous, downy pubescent perennials, 1–2 m tall, have milky sap and numerous fragrant flowers in large, umbellate clusters 5–10 cm in diameter. The widely elliptic leaves are 10–25 cm long.

These plants are native to northern and central North America and appear in our area in the northern mountains and piedmont, where they grow, often in large colonies, in meadows and along fencerows, roadsides, and forest margins. This species has been used for medicine, fiber, and food throughout its range, but the milky sap, once investigated for its potential as a source of rubber, is toxic, as is that of all *Asclepias* species, due to the presence of cardiac glycosides.

JUNE–AUGUST (157-1-10)
H-50/SEE/URR
[❍/N/$/✪/⚠]

Asclepiadaceae

PINEWOODS MILKWEED

Asclepias humistrata Walter

The sprawling, 2–7 dm long stems of this glabrous perennial have distinctive broadly ovate, opposite, sessile leaves, only slightly longer than broad, with pink to lavender veins. The 3–5 cm broad umbels arise from the axils of the upper leaves.

Growing in the sandhills and sandy pine-oak woods in the coastal states from North Carolina to Louisiana, this Milkweed is striking in its sandy habitat—which makes it a great choice for dry, sandy habitats in the wild garden. Collect some seed and try it.

MAY–JUNE (157-1-11)
H-50/SOE/URR
[❍/D/△]

Asclepiadaceae

VARIEGATED MILKWEED

Asclepias variegata Linnaeus

This attractive perennial, 2–10 dm tall, has 2 or 3 pairs of opposite, ovate leaves, 5–14 cm long and 2 to 4 compact, umbellate flower clusters, 3–6 cm in diameter.

A relatively uncommon native of the eastern United States north to New York, Variegated Milkweed grows in open deciduous woodlands and along forest margins at scattered localities, essentially throughout North Carolina.

MAY–JUNE (157-1-12)
H-50/SOE/URW
[◗/N/△]

Asclepiadaceae

CLIMBING MILKWEED

Matelea carolinensis (Jacq.) Woodson

These twining, herbaceous perennial vines with milky sap have opposite, heart-shaped leaves that are 8–14 cm long. The maroon flowers are about 2 cm across and are pollinated by flies.

Native to the southeastern United States from Maryland to Texas, these interesting perennials are infrequent in the open, deciduous woods and along stream banks in the piedmont of North Carolina. Climbing Milkweed is a good "conversation plant" for the wild garden.

APRIL–JUNE (157-3-5)
V-5O/SOE/URM
[◗/M]

Convolvulaceae

DODDER; LOVE VINE

Cuscuta rostrata Engelmann

There are 8 species of these slender, leafless, orange-stemmed, annual parasitic vines of the Morning Glory family in North Carolina. Some are parasitic on clover and other herbaceous plants, and some are parasitic on woody plants. The small, tubular flowers of all our species are usually less than 6 mm long.

This species, native to the southern Appalachians, is found on woody and semi-woody plants (often Blackberry) in our mountains.

AUGUST–SEPTEMBER (158-1-7)
V-5N/—/UTW
[❍/N/P]

Convolvulaceae

HEDGE BINDWEED

Calystegia sepium (L.) R. Br.

The freely branching, trailing stems of this perennial vine, 5–20 dm long, have triangular, sagittate leaves 5–10 cm long. The large flowers, with corolla tubes 4–7 cm long, are solitary in leaf axils and vary from white to rose-purple.

Scattered throughout North Carolina in fields, waste places, and disturbed sites, Hedge Bindweed is native throughout temperate North America but is listed as an invasive weed in many parts of its range. The root was used historically as a purgative.

MAY–AUGUST (158-6-2)
V-5A/SNL/SFW
[❍/N/✪]

Convolvulaceae

RAILROAD VINE; REDSTAR

Ipomoea coccinea L.

The bright crimson color of the 2–2.3 cm long salverform corolla of this introduced annual attracts hummingbird pollinators, assuring a good seed set in these twining annual vines. The alternate, ovate, cordate leaves, 4–9 cm long, are rarely lobed.

Found in fields and waste places scattered throughout the state and more broadly throughout the eastern and central states from New York to Texas, Railroad Vine is invasive and should not be introduced into the garden unless it is closely watched. It is easy to recognize (and destroy) in the seedling stage.

AUGUST–FROST (158-7-2)
V-5A/SCE/SFR
[❍/N]

Convolvulaceae

COMMON MORNING GLORY

Ipomoea purpurea (L.) Roth

These herbaceous, annual, twining vines, introduced into the United States as garden plants, have become thoroughly naturalized. The showy flowers, lasting only a few hours, are 3.5–5.5 cm long.

This common weed in cultivated and fallow fields, along roadsides, and in waste places throughout our state and through much of the United States is easily grown from wild-collected seed—and is easy to recognize and destroy in the seedling stage if it is unwanted.

JULY–SEPTEMBER (158-7-3)
V5A/SCE/UFB
[❍/N/$]

Convolvulaceae

MORNING GLORY

Ipomoea macrorhiza Michaux

This robust herbaceous perennial vine has pubescent, crinkled leaves (the lower ones often 3-lobed), 5–15 cm long. The large flowers, with a pubescent calyx, are 3–8 cm long.

Southeastern in distribution, this plant is infrequent to rare at its northern limit in sandy clearings and on beaches in Brunswick Co.

JUNE–JULY (158-7-10)
V5A/SOL/SFW
[❍/D]

Convolvulaceae

MAN-ROOT; MORNING GLORY

Ipomoea pandurata (L.) G. F. W. Meyer

Often several to many long, trailing, usually purple stems grow from the crown of the large, starchy root of this native perennial. The large flowers are similar to those of the seaside *I. macrorhiza*, but this species differs in its smooth leaves, glabrous calyces, and inland habitat.

Widespread and relatively common through much of the eastern and central United States, Man-Root is found throughout North Carolina on open, often dry roadsides, in old fields, and along fencerows. The Cherokee used Man-Root medicinally for treating rheumatism, asthma, and indigestion, and also ate it as a vegetable.

MAY–JULY (158-7-11)
V5A/SCE/UFW
[○/D/⊕]

Polemoniaceae

ANNUAL PHLOX

Phlox drummondii Hooker

The lower leaves of these colorful, low, 1–7 dm tall, erect annuals are opposite; the upper leaves are alternate. The flowers, with tubes 12–16 mm long, may be dark pink, white, or variegated.

This Phlox is native to Texas but has spread throughout the Southeast along sandy roadsides and in fields and waste places of the coastal plain. Annual Phlox is colorful and easy to grow and is highly recommended for the sunny wild garden.

APRIL–JULY (159-1-1)
H-50/SEE/UTR
[○/D/$]

Polemoniaceae

TRAILING PHLOX

Phlox nivalis Loddiges

This prostrate, trailing, evergreen, semiwoody perennial with deciduous flowering shoots, 1.5–3 dm tall, has linear leaves, 4–20 mm long. The flowers, with tubes 10–16 mm long, may be pink to rose or lavender, or rarely white, the eye usually dark; the petal lobes may be either entire or notched at the end.

Trailing Phlox is native to the southeastern United States north to Virginia and occurs primarily in our piedmont on sandhills, in pinelands, and in dry deciduous woods, where it sometimes forms large patches. This is a great choice for the rock garden.

MARCH–MAY (159-1-2)
H-50/SLE/UTR
[◗/D/$]

Polemoniaceae

CAROLINA PHLOX

Phlox carolina Linnaeus

The one to several flowering stems of this showy perennial, up to 10 dm tall, have opposite, ovate-lanceolate to elliptic leaves, 4–12 cm long. The compact terminal inflorescence, with individual flowers 15–25 mm long, is usually about as broad as long.

This southeastern species, ranging north to Illinois and west to Texas, is found in deciduous woods and along forest margins more or less throughout our state. It intergrades with two other tall species of Phlox, *P. maculata* and *P. glaberrima*, that have a more northern range. Carolina Phlox is often cultivated and is a great addition to the wild garden.

MAY–JULY (159-1-9)
H-50/SOE/UTR
[◗/N/$]

Hydrophyllaceae

WATER LEAF

Hydrophyllum virginianum Linnaeus

This herbaceous perennial, 3–8 dm tall, has pinnately divided leaves, 6–12 cm long, and small, rather inconspicuous flowers with the stamens exserted well beyond the 6–10 mm long corolla.

Native through much of northeastern North America, Water Leaf grows in rich woods and along stream banks in our mountains and, less frequently, in the piedmont. Native Americans used the leaves and young shoots for greens and the root for medicine.

APRIL–JUNE (160-4-3)
H-5A/PES/UCB
[●/W/$/✪]

Hydrophyllaceae

PHACELIA

Phacelia bipinnatifida Michaux

This glandular-pubescent biennial, 1–6 dm tall, bears attractive blue, bell-shaped flowers that are 10–15 mm broad. As the specific name implies, the 5–8 cm long leaves are twice pinnately divided.

More common westward in the central United States, these plants are infrequent in our mountain counties, where they grow on rocky slopes, along creek banks, and in deciduous woods. They are attractive as a solid early spring groundcover in a shade garden and are easily grown from seed. They self-seed in the garden, and thus your population of these Phacelias will usually maintain itself.

APRIL–MAY (160-5-1)
H-5A/POL/UCB
[●/M]

Hydrophyllaceae

SCORPION WEED

Phacelia purshii Buckley

These small annuals, 1–3 dm tall, but occasionally with stems reaching a length of 5 dm, have pubescent, pinnately cleft or parted leaves up to 4.5 cm long. Each inflorescence has 10 to 30 flowers with lightly fringed corollas 6–12 mm broad.

Rare in our area, these plants range primarily west of the Appalachians in the central United States and grow in moist meadows and on roadsides in only two of our mountain counties. ✗

MAY–JUNE (160-5-5)
H-5A/PEL/UCW
[◗/M/R]

Hydrophyllaceae

FRINGED PHACELIA

Phacelia fimbriata Michaux

These small, showy, 1–5 dm tall annuals are closely related to *P. purshii* but differ in that there are fewer flowers in the inflorescence of the Fringed Phacelia and, as the name implies, the petals are more deeply fringed.

Native to the southern Appalachians this *Phacelia* is known from only four of our mountain counties, where it grows, often in showy masses, along stream banks and in low woodlands. This spectacular Phacelia is well worth the effort to propagate by seed for the moist shade garden. ✗

APRIL–MAY (160-5-6)
H-5A/PEL/UCW
[●/M/$/R]

Boraginaceae

WILD COMFREY

Cynoglossum virginianum L.

This rather rank, 3–8 dm tall biennial or perennial herb has large elliptic-ovate, petiolate basal leaves, 1–2 dm long, and clasping cauline leaves. The unbranched flowering stalk bears a few-flowered raceme of blue, sometimes white, flowers.

This native is infrequent in meadows and along roadsides, primarily of our mountains and piedmont and more broadly throughout the eastern and central United States. As its name implies, this species was thought to have the medicinal qualities of Comfrey, another member of the Borage family. The Cherokee used the leaves of this plant for tobacco and the root as a cancer treatment.

APRIL–JUNE (161-2-2)
H-5A/SEE/RFB
[❍/N/$/✪]

Boraginaceae

BLUE WEED; VIPER'S BUGLOSS

Echium vulgare Linnaeus

This bristly, often weedy introduced biennial, usually 3–6 dm or more tall, has numerous circinate buds (in a tight coil) that slowly straighten as new flowers open from the base to the tip of the inflorescence. The 1–2 cm long, zygomorphic flowers turn pink with age.

Blue Weed is found in dry pastures and along roadsides in a few of our mountain and piedmont counties and throughout much of the United States. A tea made from its leaves was used as a folk medicine for pain and inflammation; the Cherokee used the leaves and roots to treat kidney disorders. The coarse hairs cause skin irritation, and all parts of the plant are mildly toxic.

JUNE–SEPTEMBER (161-4-1)
H-5A/SEE/RFB
[❍/D/✪/⚠☞]

Boraginaceae

TRUE FORGET-ME-NOT

Myosotis scorpioides Linnaeus

The linear-oblong basal leaves of this low introduced perennial, 3–7 dm tall, can be up to 8 cm long, as can the circinate flower stalk as it grows and "unrolls" the light blue flowers with a yellow center, or eye.

A native of Eurasia, this plant is now found through much of temperate North America and is frequent in bogs and along small streams in our mountains. This extremely attractive little plant is a great addition for a sunny bog garden.

MAY–AUGUST (161-7-1)
H-5A/SLE/RFB
[❍/W/$]

Boraginaceae

VIRGINIA BLUEBELLS

Mertensia virginica (L.) Persoon

This glabrous perennial, 3–7 dm tall, has petiolate basal leaves up to 2 dm long and round-tipped cauline leaves. The funnel-form flowers are in a terminal cluster and have tubes up to 2.5 cm long and a shorter flared portion known as the "limb." The blue flowers are pink in bud and again as the flower ages, presumably after pollination.

This very rare plant occurs in only one county in North Carolina, but is common in cooler areas throughout much of the eastern and central United States in rich woods and bottomlands. The Cherokee used this species for treating whooping cough and consumption. Virginia Bluebells are highly valued for the wild flower garden. ✗

MARCH–JUNE (161-10-1)
H-5A/SOE/UFB
[◗/M/$/R]

Verbenaceae

BRAZILIAN VERVAIN

Verbena brasiliensis Vellozo

The one to several sharply angled, scabrous stems of this introduced perennial are 1–2.5 m tall; the elliptic, serrate, opposite leaves are 4–10 cm long; and the numerous terminal flowering spikes are 0.5–4 cm long.

Fairly frequent in old fields and waste places of our coastal plain, this somewhat weedy, but interesting and drought tolerant plant ranges throughout the Southeast. Verbenas can be attractively clumped in the wild garden, where they add texture.

MAY–OCTOBER (162-1-1)
H-50/SES/ITB
[◯/D]

Verbenaceae

BEAUTY BERRY

Callicarpa americana Linnaeus

This 1–2.5 m tall shrub with pubescent twigs has simple, opposite leaves, 7–15 cm long. The axillary clusters of small flowers are seldom noticed. However, the compact, magenta fruit clusters, 3–4 cm in diameter, are quite showy in late summer and give the shrub its common name.

Beauty Berry, primarily a species of the southeastern United States, grows in moist, usually sandy woodlands and clearings at scattered localities, chiefly on our coastal plain. Native American tribes of the South used the leaves, roots, and stems to treat malarial fevers, and a southern folk remedy for dropsy made use of this plant. Beauty Berry is a colorful, easy-to-grow shrub for the sunny wild garden.

JUNE–JULY (162-4-1)
S-50/SOS/UCB
[◗/M/$/✪]

Lamiaceae

BLUE CURLS

Trichostema dichotomum L.

The 3–7 dm, weakly angled stems of this branched, pubescent, aromatic annual (or weak perennial) arise from a taproot. The opposite, elliptic leaves may be entire or crenate. The common name of this plant comes from the 4 long stamens that extend from the corolla tube and curve downward, a form that is adaptive for the transfer of pollen on the backs of bee pollinators.

Found in open, sandy or rocky woodlands throughout the state and more broadly throughout the eastern United States, these attractive, late summer–blooming plants may be grown for the wild garden from collected seed.

AUGUST–FROST (164-1-2)
H-50/SEE/PZB
[◗/N/$]

Lamiaceae

SKULLCAP; HELMET FLOWER

Scutellaria elliptica Muhlenberg

These perennials, unlike most other members of the Mint family, are not aromatic. The pubescent 1.5–8 dm tall stems with opposite, elliptic leaves bear attractive, strongly 2-lipped flowers, 1.2–2 cm long, in an open terminal raceme. The common name comes from the shape of the upper lobe of the calyx, which enlarges in fruit and resembles a skull or helmet.

Native through much of the southeastern and central United States north to New York, Skullcap is found throughout our state growing in mixed deciduous woodlands and on roadsides. Two other Skullcaps, *S. lateriflora* and *S. integrifolia*, are common in our area. This colorful and well-behaved plant is a good candidate for the sunny garden. Several Skullcaps have been used in traditional medicine and are important today in alternative herbal medicine.

MAY–JUNE (164-5-8)
H-50/SED/RZB
[◗/N/$/✪]

Lamiaceae

HEAL-ALL; SELF HEAL

Prunella vulgaris L.

This square-stemmed perennial weed, 1–8 dm tall, with opposite, elliptic to lanceolate, serrate leaves, produces a tight cluster of attractive 2-lipped flowers, 1–2.3 cm long.

A nonaromatic Mint widespread in fields, pastures, and lawns and along roadsides, Heal-all grows throughout temperate North America. As the name implies, this species has been used in folk and Native American medicine for a wide spectrum of ailments. It is listed as invasive in parts of its range.

APRIL–FROST (164-12-1)
H-50/SES/KZB
[❍/N/✪]

Lamiaceae

OBEDIENT PLANT

Physostegia virginiana (L.) Bentham
[*Dracocephalum virginianum* Linnaeus]

This glabrous, rhizomatous perennial is 3–10 dm tall, with 4 rows of 1.5–3 cm long, zygomorphic flowers in a terminal raceme. The common name comes from the tendency of the flowers, once pushed to one side, to remain in the new position.

Physostegia is native to the northeastern and central United States south to our area, where it grows in bogs and low meadows of the mountains and, much less frequently and more scattered, in the piedmont and coastal plain. Often cultivated, this species is somewhat aggressive and must be controlled in the garden. It makes a good cut flower as the blooms open up the stalk.

JULY–OCTOBER (164-13-1)
H-50/SES/RZR
[❍/M/$]

Lamiaceae

HENBIT

Lamium amplexicaule Linnaeus

The square stems, opposite leaves, and—on close inspection— attractive 2-lipped flowers identify this small, introduced, 1–4 dm tall annual or biennial as a member of the Mint family.

Henbit is a common weed in lawns, pastures, and cultivated fields and on roadsides throughout the state and much of temperate North America, but large populations in old fields can be quite colorful when the plants bloom.

MARCH–MAY (164-19-1)
H-50/SOD/UZB
[○/N]

Lamiaceae

HEDGE NETTLE

Stachys latidens Small

As with other members of its family, the flowers of this perennial Mint, 3–8 dm tall, are bilaterally symmetrical or zygomorphic. The coarsely toothed leaves are 7–8 cm long.

Five species of *Stachys* occur in North Carolina. This one, a native of the southern Appalachians, grows in marshes, wet meadows, and woodlands of the mountains.

JUNE–AUGUST (164-20-5)
H-50/SNS/UZR
[◗/M]

Lamiaceae

LYRE-LEAVED SAGE

Salvia lyrata Linnaeus

The pinnately lobed to divided lyrate leaves, 5–17 cm long, of this herbaceous perennial are mostly basal and form a winter rosette from which grows the 3–8 dm flowering stalk. The attractive blue 2-lipped flowers are 1.5–3 cm long.

A weedy native of the eastern and central United States north to Connecticut, this *Salvia* grows on roadsides and in lawns and meadows throughout the state. The Cherokee made used of this species medicinally to treat coughs and asthma.

APRIL–MAY (164-22-1)
H-5B/SEL/RZB
[○/N/$/⊙]

Lamiaceaae

BEE-BALM; OSWEGO TEA

Monarda didyma Linnaeus

These strong-scented, stoloniferous perennials, 7–18 dm tall, have ovate-lanceolate leaves, 8–15 cm long. The compact, crownlike whorl of 2.5–4 cm long, showy, crimson flowers, with brightly colored leaflike bracts just below, is 8–10 cm in diameter.

Native to the northeastern United States and following the mountains south to our area, these attractive plants are conspicuous but infrequent on moist road banks and open seepage slopes in the mountains. Widely planted as ornamentals, these plants make a colorful show in the wild garden, where they attract hummingbird pollinators. Native Americans used the leaf tea for treating colic, fevers, and a wide variety of other ailments, and, historically, it was used to expel worms and gas.

JULY–SEPTEMBER (164-23-1)
H-5O/SOS/UZR
[◗/M/$/⊙]

Lamiaceae

WILD BERGAMOT; BEE-BALM

Monarda fistulosa Linnaeus

The 4–12 dm tall, square stems of this rhizomatous perennial are pubescent along the angles; the opposite, ovate to lanceolate leaves are 4–11 cm long. The 2.5–2.8 cm long flowers are pink, lavender, purple, or, rarely, white.

These attractive Mints are native through much of temperate North America and grow commonly on wooded slopes, along roadsides, and in meadows of our mountains and, less frequently, in our piedmont. Wild Bergamot has been used medicinally for a large number of ailments throughout its range, and it is a great choice for the wild garden, where the blue-violet flowers attract bees.

JUNE–SEPTEMBER (164-23-3)
H-50/SOS/UZB
[◗/M/$/✪]

Lamiaceae

DOTTED HORSEMINT

Monarda punctata Linnaeus

These 3–10 dm tall perennials have opposite, lanceolate leaves, 3–9 cm long. A series of lavender or white bracts visible just below the compact whorl of yellow flowers lightly spotted with purple, add color and distinction to the 3–5 cm broad inflorescence.

Ranging through much of the eastern and central United States north to Vermont and Minnesota, Dotted Horsemint is relatively common in our coastal plain and frequent to infrequent in the piedmont. Like its relatives, this *Monarda* has an extensive history of medicinal use and was once planted as a commercial source of thymol. It grows in sandy or rocky fields and open woods and is a colorful addition to the sandy wild garden.

AUGUST–SEPTEMBER (164-23-4)
H-50/SNS/UZR
[❍/D/$/✪]

Lamiaceae

HOARY MOUNTAIN MINT

Pycnanthemum incanum (L.) Michaux

This pungent, aromatic, rhizomatous perennial, usually with several erect stems, 1–2 m tall, from a single root crown, has ovate, lanceolate to elliptic leaves, 5–14 cm long, that are whitened near the inflorescence. The small flowers are in compact, 1.3–2 cm broad, heads that are terminal and in the upper leaf axils.

Plants of the eastern United States, 8 relatively similar species of *Pycnanthemum* are lumped under the common name Mountain Mint. These plants grow in woodlands, thickets, pastures, and old fields throughout our state. The leaves were widely used in tea taken for headache and as a cold remedy. All of these plants do well in the wild garden in light shade.

JUNE–AUGUST (164-28-8)
H-50/SOS/UZB
[◗/N/$/✪]

Lamiaceae

PEPPERMINT

Mentha piperita L.

The square, purplish, erect or prostrate stems of these naturalized, aromatic, rhizomatous perennials, 6–16 dm tall, root readily in water or moist soil and often form large clumps. The small, 2-lipped flowers are in loose or interrupted terminal clusters.

Found in marshes, ditches, and wet meadows scattered throughout the state and much of the United States, Peppermint has a long history of medicinal use in Europe and America.

JUNE–FROST (164-32-7)
H-50/SNS/PZB
[❍/M/$/✪]

Lamiaceae

HORSE-BALM; STONEROOT

Collinsonia canadensis Linnaeus

The opposite, ovate, coarsely toothed leaves, the panicle of yellow flowered racemes, and the characteristic lemon scent serve to identify this 4–15 dm tall perennial from a distance. Notice, on close inspection, that the lower lip of the yellow flower is fringed.

An interesting Mint that ranges broadly over much of temperate eastern North America, Horse-Balm grows in rich, moist woods, primarily in our mountains and piedmont. These plants would make an interesting addition to the shade garden for their form and fragrance. The roots and leaves have diuretic properties associated with their alkaloid content, and there are possibilities for the use of Horse-Balm as an anti-oxidant.

JULY–SEPTEMBER (164-33-1)
H-5O/SOS/PZY
[●/M/✪]

Solanaceae

APPLE OF PERU

Nicandra physalodes (L.) Persoon

This glabrous annual is 2–10 dm tall, with succulent stems and a 3–4 cm broad corolla of 5 fused petals. The fleshy green berry, 1.3–2 cm in diameter (a cousin of the Mexican "tomatillo"), is hidden by the 5-angled, inflated calyx, which looks like a miniature Japanese lantern.

Introduced from South America, these weedy plants are now naturalized throughout the eastern and central United States and, in North Carolina, are found in fields and waste places at scattered localities, chiefly in the mountains. These plants are easy to grow from seed.

JULY–SEPTEMBER (165-2-1)
H-5A/SND/SRB
[❍/N]

Solanaceae

HORSE NETTLE

Solanum carolinense Linnaeus

This prickly, rhizomatous perennial has simple or weakly branched stems, 2–8 dm tall, with lavender to white, 5-parted flowers about 2–3 cm broad. The globular, dull yellow berries are 1–1.5 cm in diameter.

A weedy native through much of the United States, Horse Nettle grows in old fields, gardens, and waste places and on roadsides throughout the state. Both leaves and berries have been used for medicine, but note that the berries are toxic; do not ingest!

MAY–JULY (165-5-5)
H-5A/SOL/RRW
[❍/D/✪/◉]

Solanaceae

JIMSON WEED

Datura stramonium Linnaeus

The stems of this rank-smelling, glabrous annual from a short taproot are 5–15 dm tall and are often marked with purple. The lanceolate, irregularly incised leaves are 7–15 cm long. The white to lavender corolla is 7–10 cm long, and the distinctive, ovoid capsule is spiny.

This common weed throughout most of North America is found throughout our area on roadsides and in fields, barn lots, and waste places. This species has numerous alkaloids with medicinal benefit—currently for eye diseases and Parkinson's disease—and has historically had use as a folk remedy for cancer. All parts of this plant, but primarily the leaves and the black seeds, are toxic and cause hallucinations.

JULY–SEPTEMBER (165-8-1)
H-5A/SNL/SFW
[❍/N/✪/◉]

Scrophulariaceae

PRINCESS TREE

Paulownia tomentosa (Thunberg) Steudel

The opposite, ovate, entire 1.5–3 dm long leaves of this fast-growing introduced Asian tree, up to 15 m tall, are pubescent above and tomentose beneath and come out in spring after the large panicles of fragrant, purple, 2-lipped, 5–6 cm long flowers. The numerous minute, winged seeds contained in each large, woody capsule are widely dispersed by the wind. The brown spikes of next spring's flower buds appear when the leaves fall in the autumn.

Escaped from cultivation in scattered localities in our state and more generally in the Eastern United States from Massachusetts to Texas, these trees have become quite common, and noticeable, along many of North Carolina's highways, especially in the mountains. Princess Tree is easily cultivated, though somewhat invasive. It has a long history of medicinal use in China, where it is native, and the wood is highly prized in Japan for making small boxes.

APRIL–MAY (166-1-1)
T-5A/SOE/PZB
[❍/N/$/✪]

Scrophulariaceae

MONKEY FLOWER

Mimulus ringens Linnaeus

The square stem of this glabrous native perennial is 8–10 dm tall, but, unlike in the closely related *Mimulus alatus*, the angles are not winged. The strongly 2-lipped lavender flowers of both species are 2.5–4 cm long.

The primary range for *M. ringens* is throughout much of the eastern and central United States and southern Canada south to Louisiana. In our area Monkey Flower occurs chiefly in the mountains and upper piedmont, where it grows in marshes, bogs, wet meadows, and along stream banks. Easily grown from seed, these plants make attractive additions to the sunny wild garden.

JUNE–SEPTEMBER (166-11-2)
H-50/SNS/SZB
[❍/W/$/✪]

Scrophulariaceae

WOOLY MULLEIN; MOTH MULLEIN

Verbascum thapsus Linnaeus

These rank, wooly-pubescent, yellowish-green biennials produce flower stalks 1–2 m tall; the wooly elliptic leaves are 10–30 cm or more long. The very fragrant flowers, 15–22 mm broad and weakly zygomorphic, are in clusters of 3 in a compact elongate inflorescence; the center flower of each cluster opens first.

This introduced European weed is common along roadsides and in fallow fields, pastures, and waste places, essentially throughout the state and throughout North America. These plants have a long history of medicinal use in both Europe and North America. This, and other glabrous species of *Verbascum*, all of which are easy to grow, have come into cultivation in the last decade, and seeds of many new horticultural hybrids of various colors are now available.

JUNE–SEPTEMBER (166-12-3)
H-5A/SEE/IFY
[○/D/$/✪]

Scrophulariaceae

TURTLEHEAD

Chelone cuthbertii Small

The very strongly 4-ranked flowers and the narrowly lanceolate, opposite, 7–12 cm long, sessile leaves of this native perennial are quite distinctive and separate this species from other Turtlehead species found in North Carolina. The terminal spikes of 3–4 cm long, zygomorphic flowers are at the end of slender stems 5–10 dm tall.

Narrowly distributed in Virginia, the Carolinas, and Georgia, these handsome plants grow along stream banks and in wet meadows or bogs, chiefly in our mountains. A great possibility for the moist, mountain wild garden, this more rare of the Turtleheads should be propagated by seed.

JULY–SEPTEMBER (166-13-1)
H-5O/SNS/IZB
[◗/W/$]

Scrophulariaceae

WHITE TURTLEHEAD

Chelone glabra Linnaeus

The erect or spreading, 6–15 dm long stems of this native perennial have opposite, elliptic-lanceolate, petiolate leaves, 10–18 cm long. The 3–4 cm long, 2-lipped flowers are not arranged in 4 strict ranks.

Native to northeastern and central North America south to Georgia, White Turtlehead grows in low, moist meadows and thickets at widely scattered localities in all three provinces of our state but is nowhere common. Relatively easily propagated by seed and cuttings, these plants are highly recommended for the wet shade garden. White Turtlehead was used in Native American and folk medicine to stimulate appetite and to reduce fever and jaundice.

AUGUST–OCTOBER (166-13-4)
H-50/SES/IZW
[◗/W/$/✪]

Scrophulariaceae

BEARD TONGUE; PENSTEMON

Penstemon canescens Britton

This erect, pubescent, often sticky native perennial has flower stalks 3–7 dm tall. The 2-lipped flowers, 24–32 mm long, have one sterile bearded stamen, giving rise to the common name.

Native to the Appalachians from Pennsylvania to Georgia and west to Illinois, these plants grow on rocky banks and roadsides and in meadows in the mountains and a few counties in the upper piedmont. There are 6 Penstemons in North Carolina that can be grown from collected seed, and one or more would be well worth planting in the native garden.

MAY–JULY (166-14-2)
H-50/SNS/PZB
[❍/N/$]

Scrophulariaceae

TOAD FLAX

Nuttallanthus texanus (Scheele) D. A. Sutton
[*Linaria canadensis* var. *texana* (Scheele) Pennell]

The erect 1–7 dm flowering stems of this annual or biennial arise from a basal rosette of short, prostrate, sterile stems. The linear leaves, 3–20 mm long, are alternate on the flowering stems and opposite to whorled on the prostrate stems. The strongly 2-lipped flowers, 10–15 mm long, are in a raceme.

Native to the Southeast and much of the central and western United States, these plants are found in fallow fields (frequently growing in colorful masses with *Rumex*) throughout our coastal plain and piedmont. The smaller-flowered *N. canadensis* is also common in our area.

MARCH–MAY (166-16-1)
H-5A/SLE/RZB
[○/N]

Scrophulariaceae

SPEEDWELL; VERONICA

Veronica persica Poiret

The flowers of this 1–4 dm tall introduced annual are 7–12 mm broad and, unlike the flowers of most other members of its family, they have only 4 petals, the upper 2 of the usual 5 petals being completely fused. The corolla is therefore only slightly zygomorphic, with a very short tube.

This colorful little weed is found in lawns and along roadsides at numerous scattered localities in our mountains and piedmont and throughout much of North America. One or another of the 11 species of *Veronica* in North Carolina can be common in early spring. Plants of several species of *Veronica* have had medicinal use.

MARCH–JUNE (166-20-5)
H-40/SOD/SZB
[○/N]

Scrophulariaceae

FALSE FOXGLOVE

Aureolaria pedicularia (L.) Rafinesque

These spreading, branched, glandular-pubescent annuals, 3–10 dm tall, with yellow, tubular flowers, 3–4.5 cm long, are semiparasitic on the roots of trees in the Black Oak group. The elliptic to lanceolate, entire leaves of the basal rosette are 4–8 cm long.

A native of the eastern United States south to Georgia, this relatively rare *Aureolaria* grows in sandy woodlands and dry, open, deciduous forests locally throughout the state. These attractive plants may be started from seed in the shady garden if their Black Oak host plants are present. ✗

SEPTEMBER–OCTOBER (166-24-1)
H-50/SEE/SZY
[◗/D/✪/P]

Scrophulariaceae

DOWNY FALSE FOXGLOVE

Aureolaria virginica (L.) Pennell

The erect, unbranched, pubescent flowering stems of this native perennial, 3–10 dm tall, have lanceolate leaves, 6–12 cm long, and a terminal raceme of 3.5–4.5 cm long tubular yellow flowers.

Distributed throughout the eastern United States north to New Hampshire, these perennials are semiparasitic on the roots of trees of the White Oak group. They grow in or along the margins of deciduous woodlands over most of the state. They are pollinated by bumblebees that turn upside down as they enter the large flowers. ✗

MAY–JULY (166-24-3)
H-50/SNE/RZY
[◗/D/✪/P]

Scrophulariaceae

GERARDIA

Agalinis purpurea (L.) Pennell

These annuals, semiparasitic on the roots of grasses, are often profusely branched and are 1–12 dm tall, with linear leaves, 1–4 cm long. The pink flowers are 2–4 cm long and nearly as broad, and are yellow-lined, purple-spotted, and hairy at the base of the corolla lobes.

A colorful, sporadic weed along low roadsides, it is native to pond margins, roadsides, and low meadows of the eastern and central United States, and occurs locally in all provinces of North Carolina. Ten other species of *Agalinis* are found in our state, mostly in the coastal plain. Collect some seed and scatter it in a sunny corner of your lawn; the plants may thrive and self-seed. ✗

AUGUST–FROST (166-25-3)
H-50/SLE/RZR
[❍/M/P]

Scrophulariaceae

INDIAN PAINT BRUSH

Castilleja coccinea (L.) Sprengel

The crimson bracts, or modified leaves, around the flowers of this pubescent, native, 2–6 dm tall annual or biennial are cleft into 3 to 5 segments and are more brilliantly colored than the slender, 2.5 cm long yellow flowers. The spike is 4–6 cm long in flower, expanding to 1–2 dm in fruit.

Found throughout much of eastern and central North America, plants of this species are parasitic on the roots of grasses and grow in moist meadows and along woodland margins, chiefly in our mountains and, less frequently, in the piedmont. Because Indian Paint Brush is not too picky about its host, it is a possibility for propagation by seed in the wild garden (or a sunny spot in your lawn) where the crimson bracts and tubular flowers will attract hummingbirds. The genus is much larger in the western United States. ✗

APRIL–MAY (166-28-1)
H-4A/SEL/ITR
[❍/M/$/✪/P]

Scrophulariaceae

WOOD BETONY; LOUSEWORT

Pedicularis canadensis Linnaeus

These pubescent perennials, 1–4 dm tall, have yellow or yellow and red flowers, 18–22 mm long, in a compact spike 3–4 cm in diameter. The alternate, fernlike, elliptic leaves of the flowering stalk are 5–15 cm long.

Relatively infrequent in moist, deciduous woodlands of the mountains and piedmont of North Carolina, Wood Betony is found throughout much of eastern and central North America west to Colorado. A fall-flowering species, *P. lanceolata*, with opposite stem leaves, occurs in our southwestern mountains. Worldwide, there are some 600 species of *Pedicularis*, the largest genus in the family Scrophulariaceae, all pollinated by bumblebees. This species has a long list of medicinal uses by Native Americans throughout its range.

MAY–JULY (166-29-1)
H-5A/SEC/IZY
[◗/M/⊕]

Bignoniaceae

CROSS VINE

Bignonia capreolata Linnaeus
[*Anisostichus capreolata* (L.) Bureau]

This woody vine with large, colorful flowers, 4–5 cm long, has opposite, pinnate leaves that are reduced to two 6–15 cm long leaflets, and a terminal tendril. The tendrils cling to bark and enable Cross Vine to climb to the tops of tall trees, where it often blooms profusely.

Often found growing in alluvial forests and deciduous woods, chiefly of our piedmont and coastal plain, these handsome plants are native to the southeastern United States west to Texas and north to Illinois. The numerous winged, wind-dispersed seeds scatter widely and germinate readily; thus, these vines can be considered aggressive. When a little care is taken to control their spread, however, they make an attractive addition to the wild flower garden fence.

APRIL–MAY (167-1-1)
V-50/PNE/UZR
[○/M/$]

Bignoniaceae

TRUMPET CREEPER; COW-ITCH VINE

Campsis radicans (L.) Seemann

This native woody vine has opposite, pinnately compound leaves with 7 to 15 leaflets, each 3–4 cm long. The showy, trumpet-shaped, 6–8 cm long flowers are red to bright orange. The pendant, crescent-shaped fruit, 1–1.8 dm long, has small nectaries that attract ants, which presumably protect the green fruit from predators before the mature capsule releases its small, wind-dispersed seed.

Native to the southeastern United States from New Jersey to Texas, Trumpet Creeper has escaped from cultivation further north and west and grows along fencerows and the margins of low woods and thickets throughout our area. The leaves and flowers are toxic if eaten and can irritate the skin.

JUNE–JULY (167-2-1)
V-50/POS/UZR
[◯/N/$/⚠☞]

Bignoniaceae

INDIAN CIGAR; CATAWBA TREE

Catalpa speciosa Warder ex Engelmann

The long, slender, fibrous fruits of this tree were sometimes smoked by rural children, which may account for one of the common names. Large caterpillars that feed on the leaves account for a less common colloquial name, Fish-Bait Tree. The large, attractive white flowers, 4 cm or more long, are variously marked with yellow and purple in the center. The opposite, ovate leaves are 1–3 dm long and about as wide.

Native to the central United States, Indian Cigar is widely planted, and often escaped, in much of the eastern United States. It grows on road banks and in waste places at a few localities in our piedmont and mountains and is often planted as an ornamental tree that can grow up to 30 m tall.

MAY–JUNE (167-3-2)
T-50/SOE/PZW
[◯/M/$]

Orobanchaceae

BEECH-DROPS

Epifagus virginiana (L.) Barton

The slender brownish, purple, or yellow, branched stems of this annual, obligate parasite on the roots of beech trees are 1–5 dm tall. The racemes have small, sterile, brownish, chasmogamous flowers above and fertile cleistogamous flowers below.

Found in rich beech woods, primarily in our mountains and piedmont and throughout eastern North America wherever beech is found. The plant has been used as a folk remedy for cancer and is sometimes called Cancer-Root. ✗

SEPTEMBER–NOVEMBER (169-1-1)
H-4N/—/SZM
[●/M/✪/P]

Orobanchaceae

SQUAW-ROOT; CANCER-ROOT

Conopholis americana (L.) Wallroth

The numerous thick, yellowish-brown emergent stems of this parasitic perennial herb, generally 1–2 dm tall, often form large clumps on the roots of oak and various other host trees. The leaves are reduced to small brown scales. Note that the unseen cleistogamous, or closed, flowers are on nonemergent stems, self-pollinate, and produce seed underground.

Native to eastern North America, Squaw-Root occurs in dry oak forests at scattered localities, chiefly in our mountains. ✗

MARCH–JUNE (169-2-1)
H-4N/—/SZY
[●/D/P]

Orobanchaceae

BROOMRAPE

Orobanche uniflora L.

These bluish-white, glandular-pubescent annual parasites on the roots of several woody species have scapes 7–14 cm tall, with leaves reduced to brown scales, densely overlapping at the base of the flowering stalk.

This interesting parasite is found in rich woods at scattered localities in our mountains and piedmont and is widely distributed throughout North America. ✗

APRIL–MAY (169-3-1)
H-5B/—/STB
[●/M/P]

Lentibulariaceae

BUTTERWORT

Pinguicula pumila Michaux

The tiny insects that are trapped and digested by the sticky hairs on the 1–3 cm long leaves of this low (0.5–1.5 dm tall) perennial provide it with needed nitrogen, so the plant is considered "carnivorous," as are other members of this interesting family. The flowers are about 12 mm long, including the slender spur.

A rare native of southeastern coastal states from North Carolina to Texas, these plants are known from only three of our coastal counties, where they grow in low pinelands and savannas, and they are are on the state's protected list. These intricate little plants are worth propagating by seed for the moist wild garden. ✗

APRIL–MAY (170-1-1)
H-5B/SOE/SZB
[○/M/$/✪/SR]

Lentibulariaceae

YELLOW BUTTERWORT

Pinguicula lutea Walter

The large, 2–3.5 cm long, yellow corolla on scapes 1–5 dm tall and the larger, 1–6.5 cm long, viscid leaves of this perennial easily separate it from *P. pumila*. The other species in North Carolina, *P. caerulea*, also has longer corolla tubes, 2.5–4 cm.

Found in low pinelands and savannas north to Pender County in our state and south through the coastal plain to Louisiana, Yellow Butterwort is a prize for the wet, sunny, wild flower garden; grow it on moist sphagnum from collected seed.

APRIL–MAY (170-1-2)
H-5B/SOE/SZY
[❍/W/$/✪]

Lentibulariaceae

BLADDERWORT

Utricularia purpurea Walter

These floating annual or perennial aquatics have submersed stems up to 1 m long with whorls of 5 to 7 leaves that are branched into fine filaments, many with a terminal bladder that catches tiny aquatic animals. The flowering scapes, 3–15 cm long, have 1 to 3 2-lipped flowers, 9–12 mm long.

This Bladderwort is native in quiet water from southern Canada south through the coastal states to the Gulf of Mexico. In our area, it is found in shallow pools, ponds, and ditches in a few counties of the coastal plain.

MAY–SEPTEMBER (170-2-4)
A-5W/PLE/RZB
[❍/A]

Lentibulariaceae

BLADDERWORT

Utricularia inflata Walter

Each 4–20 cm long flowering scape of this native perennial aquatic arises from a whorl of inflated leaves that forms a floating platform. Each scape has 3 to 15 flowers, 1.5–2.5 cm long. The highly divided submersed leaves have many small bladders that trap equally small larvae and fish.

These plants of the southeastern United States are found north along the coast to New York and are irregularly distributed in ponds, pools, and roadside ditches at a few widely scattered localities in our coastal plain.

MAY–NOVEMBER (170-2-5)
A-5W/PLL/RZY
[○/A]

Acanthaceae

RUELLIA

Ruellia caroliniensis (Walter) Steudel

This pubescent native perennial, 1–6 dm tall, has opposite, ovate to elliptic leaves, 4–12 cm long. The funnel-form flowers, 2.5–4.5 cm long, are found in sessile clusters at the middle and upper nodes.

Native to the Southeastern United States north to Pennsylvania and Illinois, Ruellia is found in dry woods, sandy fields, and rock crevices throughout North Carolina. Ruellia is easily propagated from seed and is a great choice for dry, sunny sites in the wild garden.

MAY–SEPTEMBER (171-3-7)
H-50/SOE/KFB
[◗/D/$]

Plantaginaceae

LARGE-BRACTED PLANTAIN; BUCKHORN

Plantago aristata Michaux

Typical of Plantain species, the 1–2.5 dm long flowering scapes of this winter annual arise directly from the whorl of linear, 1–20 cm long, basal leaves. Unusual in this species are the linear, green bracts throughout the densely flowered spike that give this species its distinctive appearance.

Found on roadsides and in waste places, pastures, fields, and lawns throughout our state, where it has naturalized from its original range in the central United States, Buckhorn is now widespread in eastern North America and much of the continental United States. This is one of many Plantain species in our area that are mostly weeds of waste areas. The Cherokee used the leaves and roots of this plantain for treating poisonous bites, burns, and ulcers.

APRIL–JULY (172-1-5)
H-4B/SLE/ITG
[❍/D/✪]

Rubiaceae

BUTTON BUSH

Cephalanthus occidentalis Linnaeus

The compact, globose inflorescence of this 1–3 m tall shrub is 2–3.5 cm in diameter and is made up of many small tubular flowers, 6–10 mm long, with exserted anthers. The opposite, usually glabrous leaves are mostly 5–10 cm long.

Native through much of eastern and central North America, Button Bush grows along streams, by ponds and lakes, and in low open areas over most of North Carolina. Button Bush is a great choice for landscaping in moist, sunny areas and has a long history of use in medicine by Native Americans throughout its range.

JUNE–AUGUST (173-2-1)
S-40/SOE/KTW
[❍/M/$/✪]

Rubiaceae

PARTRIDGE BERRY

Mitchella repens Linnaeus

The small, fragrant, 9–14 mm long, trumpet-shaped flowers of this prostrate, creeping perennial always occur in pairs. Later, the 2 developing ovularies fuse to form a single berrylike fruit 7–10 mm broad. The opposite, ovate leaves are 8–20 mm long on stems 1–3 dm long.

A native of eastern North America, *Mitchella* is found in low deciduous woods more or less throughout our state and makes a wonderful groundcover for the shade garden. Native Americans used Partridge Berry throughout its range, and a tea made from its dried or fresh leaves or berries was esteemed as a wash for swellings, hives, and rheumatism.

MAY–JUNE (173-6-1)
H-40/SOE/UTW
[●/M/$/⊕]

Rubiaceae

BLUETS

Houstonia caerulea Linnaeus

These delicate but colorful, small native perennials grow in clumps and spread by slender rhizomes. The thin flower stalks are 5–10 cm tall, and the small, ovate, mostly basal leaves are 3–8 mm long and 2–5 mm wide. The tiny tubular flowers, 4–7 mm long, have either long styles and short stamens or short styles and long stamens that aid in efficient cross pollination by small beeflies that forage for nectar.

Found throughout northeastern and central North America south to Georgia and Louisiana, Bluets are common in lawns, clearings, and forest margins throughout the state, though infrequent in the coastal plain. Since Bluets are rhizomatous perennials, they can be expected to survive transplanting to the sunny wild garden. The Cherokee used a Bluet leaf tea to prevent bed-wetting.

APRIL–MAY (173-8-1)
H-4B/SOE/STB
[◗/M/$/⊕]

Rubiaceae

HOUSTONIA

Houstonia purpurea Linnaeus

Much less showy than Bluets, these native, perennial herbs have one to several upright leafy stems, 1–3 dm tall, from a single root crown. The leaves, 1–6 cm long, may be glabrous or densely pubescent. The tubular flowers, 5–7 mm long, are in cymes that are terminal on the stem or in the upper leaf axils.

Found in much of the eastern and central United States, this *Houstonia* grows in deciduous woods, primarily in our mountains and piedmont and is a candidate for the wild garden.

MAY–JUNE (173-8-5)
H-4O/SNE/UTR
[◗/N]

Rubiaceae

BEDSTRAW; STICKYWILLY

Galium aparine L.

Anyone who has walked through a mat of this prostrate or trailing, sticky-stemmed annual will remember the experience. The 2–10 dm long minutely barbed stems are angled, and the angles are armed with short, stiff, retrorse bristles. The whorled leaves, 6 to 8 per node, are 2–4.5 cm long. The tiny white flowers are born in sometimes branched, 3- to 5-flowered cymes at the nodes.

Stickywilly is common along roadsides and in meadows, woodlands, and waste places throughout the state and ranges broadly throughout temperate North America, everywhere listed as an invasive weed. A tea of the whole plant was once used as a diuretic.

APRIL–MAY (173-10-6)
H-4W/SLE/UCW
[○/N/✪]

Caprifoliaceae

BUSH HONEYSUCKLE

Diervilla sessilifolia Buckley

This low shrub, 0.5–2 m tall, has simple, opposite leaves, 5–18 cm long, that, as the specific name implies, are usually sessile, at least near the ends of the branches. The zygomorphic flowers are 1.5–2 cm long.

These plants, growing on bluffs and road banks in only a few of our mountain counties, are native to a rather restricted area of six states in the southern Appalachians. This is an attractive possibility for the wild garden in cooler parts of North Carolina.

JUNE–AUGUST (174-1-2)
S-50/SNE/UZY
[◗/N/$]

Caprifoliaceae

JAPANESE HONEYSUCKLE

Lonicera japonica Thunberg

This twining, invasive woody vine, introduced long ago from Japan as an ornamental, is often a bad weed, strangling or shading native vegetation. The wonderfully fragrant 2-lipped flowers, 3–5 cm long, may occasionally justify its existence.

Japanese Honeysuckle is common in waste places and low woodlands throughout North Carolina and much of the eastern and southern United States and is a seriously invasive weed. The bark, flowers, and leaves were used to treat a variety of ailments throughout its native range in Japan and Asia. Potential use of Honeysuckle in modern medicine suggests possible controls for this alien species. Warning: the berries are toxic; do not ingest!

APRIL–JUNE (174-2-4)
V-50/SOE/UZW
[◗/N/✪/△]

Caprifoliaceae

CORAL HONEYSUCKLE

Lonicera sempervirens Linnaeus

A glabrous, twining vine, up to about 5 m long, this Honeysuckle has opposite leaves, 3–7 cm long, the pair just below the inflorescence fused around the stem. The united petals form a slender trumpet-shaped flower, 3–5 cm long, that is attractive to hummingbird pollinators.

These perennials, sometimes cultivated, are native to much of the eastern and central United States and grow in clearings and along the margins of deciduous woodlands in our coastal plain and piedmont. Coral Honeysuckle is an excellent plant for the wild flower garden, making a colorful display either alone or with Yellow Jessamine. A recently discovered variant blooms sporadically throughout the growing season.

MARCH–MAY (174-2-5)
V-50/SOE/UTR
[◗/N/$]

Caprifoliaceae

HOBBLE-BUSH; MOOSEWOOD

Viburnum lantanoides Michaux
[*Viburnum alnifolium* Marshall]

This loosely branched, deciduous native shrub, up to 5 m tall, has ovate leaves, 1–2.5 dm long. The showy outer flowers of the umbellate inflorescence are 2.5–3 cm across and are sterile; the smaller flowers in the center are fertile.

This is a northeastern species, ranging from southern Canada to the mountains of Georgia and is found in our area only in the higher mountains, where it grows in rich coves and along stream banks. The brilliant red fruits ripen in August. This is one of many species of *Viburnum* in cultivation.

APRIL–JUNE (174-5-1)
S-50/SOS/URW
[◗/M/$]

Caprifoliaceae

POSSUMHAW

Viburnum nudum L.

This medium-sized shrub, up to 4 m tall, has opposite, elliptic leaves, 5–12 cm long. The small, ill-scented flowers are arranged in a flat umbellate inflorescence 5–10 cm across; the fleshy, blue-black fruits are 6–8 mm long.

Possumhaw grows in bogs, pocosins, bays, and low woods, and ranges throughout North Carolina and the eastern United States west to Texas. Witherod, *Viburnum cassinoides*, primarily of our mountains, is now included as a variety of *V. nudum*. Native Americans used a Possumhaw bark tea as a diuretic and tonic. Like many *Viburnum* species, Possumhaw is easy to grow and can be appealing for landscape use and for attracting birds.

APRIL–MAY (174-5-3)
S-50/SEE/URW
[◗/N/$/✪]

Caprifoliaceae

MAPLE-LEAVED VIBURNUM; ARROW-WOOD

Viburnum acerifolium Linnaeus

The opposite, deciduous, 3-lobed leaves of this 1–2 m shrub are 7–13 cm long, and their similarity to maple leaves accounts for both the scientific name and one of the common names of this species of *Viburnum*. The flat inflorescence, and later the cluster of black fruits, is 5–10 cm broad.

Native to eastern North America west to Texas, Arrow-Wood grows in deciduous forests of the mountains and piedmont. Several eastern tribes of Native Americans used the bark of *V. acerifolium* as an emetic and diaphoretic and as a remedy for cramps and colic. Like other Viburnums, this attractive shrub is a good candidate for landscape use.

APRIL–JUNE (174-5-9)
S-50/SOL/URW
[◗/M/$/✪]

Caprifoliaceae

ELDERBERRY

Sambucus canadensis Linnaeus

The soft, pithy, branching stems of this native shrub grow 1–3 m tall. The compound leaves have petioles that are 3–10 cm long and 5–11 lanceolate leaflets, 5–15 cm long. The flat inflorescence of small, fragrant white flowers, is 5–15 cm broad.

These shrubs grow in moist rich soil in open pastures, in alluvial woods, and along fencerows and forest margins throughout North Carolina and much of eastern and central North America. The fruit ripens in July or August and is attractive to birds and wildlife—as well as people who use the juicy berries for wine, jam, and jelly. There are many recorded uses of this plant by Native Americans and in folk medicine as well. Warning: the bark, roots, leaves, and unripe fruits are toxic when eaten.

MAY–JULY (174-6-1)
S-50/PNS/URW
[❍/M/$/✪/△]

Caprifoliaceae

RED ELDER

Sambucus pubens Michaux

The brown rather than white pith, the conical inflorescence of inodorous flowers, and the bright red berries of this 1–3 m shrub separate it from the purple-fruited *S. canadensis*.

Found throughout northern North America, moving south in the mountains in both the eastern and western states, Red Elder is found in our area only in the higher mountains, where it grows in, or along the margins of, deciduous or spruce-fir forests. For the right growing zone, this is a wonderful landscape shrub. Like those of the more common Elderberry, the berries are used in jams and jellies, and all parts of the plant except the ripe fruits are toxic; do not ingest.

MAY–JUNE (174-6-2)
S-50/PNS/URW
[◗/M/$/△]

Cucurbitaceae

CREEPING CUCUMBER

Melothria pendula L.

This slender, climbing, annual or perennial vine, 1–2 m long, has palmately 3- to 5-lobed, 2–8 cm long leaves typical of the cucumber family. Simple tendrils are borne to the side of the leaves, and the solitary, axillary flowers may be perfect or female. Male flowers are borne separately on the vine in long pedunculate cymes. The globose to ellipsoid, green or black berry is 5–10 cm long.

Native on roadsides and in alluvial woods, thickets, marshes, and fields, primarily in our coastal plain and piedmont, Creeping Cucumber ranges from Pennsylvania and Illinois south to northern Mexico. The berries have a strong laxative effect and are considered toxic; do not ingest!

JUNE–FROST (177-6-1)
V-5A/SOL/SCY
[◗/N/✪/△]

Campanulaceae

VENUS' LOOKING-GLASS

Triodanis perfoliata (L.) Nieuwand
[*Specularia perfoliata* (L.) A. DeCandolle]

The 1–10 dm tall flowering stems of these annuals are more or less encircled by the widely heart-shaped leaves. The open-pollinated flowers are 1–1.5 cm broad; smaller flowers that are lowest and earliest on the stem are cleistogamous and self-pollinated.

These common, somewhat weedy natives throughout most of North America grow in plowed fields, in waste places, and along roadsides throughout our area. Grow a few plants from seed for your wild garden.

APRIL–JUNE (178-1-1)
H-5A/SCS/SRB
[○/N]

Campanulaceae

BELLFLOWER

Campanulastrum americanum (L.) Small
[*Campanula americana* Linnaeus]

These weakly branched, 0.5–2 m tall annuals have lanceolate leaves, 6–17 cm long, and produce flowers in a raceme with leafy bracts in at least the lower third. The rotate flowers are 2–3 cm broad with the long style curved abruptly upward near its tip.

Bellflower ranges through the eastern and central United States north to New York and Minnesota. In North Carolina it is frequent on moist banks and along woodland margins in the mountains and much less frequent, or rare, in the piedmont. Native Americans used a tea made from its leaves to treat coughs and tuberculosis.

JUNE–SEPTEMBER (178-2-1)
H-5A/SNS/RRB
[◗/M/$/☉]

Campanulaceae

SMALL BELLFLOWER

Campanula divaricata Michaux

This Bellflower's delicate, 7–9 mm long, blue flowers, distinguished by their curved styles, are organized into diffuse panicles that are difficult to spot from a distance. Usually clumped with several to many flowering stems, 2–5 dm tall, these perennials have lanceolate to elliptic, 3–7 cm long leaves.

Common in rocky woods, on cliffs, and on talus slopes of our mountains and less frequent in the piedmont, Small Bellflower ranges from western Maryland and eastern Kentucky to Georgia and Alabama. Because of their delicate texture and tendency to clump, these plants would make a nice addition to the wild garden.

JULY–OCTOBER (178-2-2)
H-5A/SNS/PFB
[◗/N/$/☉]

Campanulaceae

CARDINAL FLOWER

Lobelia cardinalis Linnaeus

There are few, if any, more brilliant and intensely colored flowers than those of this 5–15 dm tall herbaceous perennial. The flowers, 3–4 cm long, are strongly 2-lipped, or bilaterally symmetrical, even though other members of the family have radial flowers.

Cardinal Flower grows in moist, open meadows and along stream banks in scattered localities throughout North Carolina and the eastern, central, and southwestern United States. These attractive plants, pollinated by hummingbirds, are quite easy to grow from seed and mass in the bog garden or along pond edges. Native Americans used Cardinal Flower leaf and root teas to treat fever, colds, stomachache, and other ailments; note, however, that all parts of the plant are toxic when ingested.

JULY–OCTOBER (178-6-1)
H-5A/SES/RZR
[◗/W/$/✪/△]

Campanulaceae

GREAT LOBELIA

Lobelia siphilitica Linnaeus

The unbranched stems of this coarse perennial grow to 1 m or more tall, with alternate, elliptic to lanceolate leaves, 5–25 cm long. The 12–15 mm long, 2-lipped flowers are arranged in a terminal, open raceme up to 6 dm long.

Native to the northern and central United States, these plants are found only in the western mountains of North Carolina where they grow in wet meadows, in low woods, and along stream banks. This colorful and adaptable plant may be grown with Cardinal Flower to create a lovely display, but watch out for purple-flowered hybrids. Native Americans used the leaves and roots to treat fever, coughs, and headaches, among other ailments. As with other Lobelias, all parts of the plant are toxic when ingested.

JULY–OCTOBER (178-6-2)
H-5A/SES/RZB
[◗/M/$/✪/△]

Campanulaceae

INDIAN TOBACCO

Lobelia inflata Linnaeus

These pubescent native annuals, 1–12 dm tall, have lanceolate to ovate leaves, 3–11 cm long, and racemes, up to 2.5 dm long, with small white to pale blue flowers, the floral tubes 2–4 mm long. The papery seed pods are inflated—thus the species name.

Found in fields, meadows, woodlands, and gardens throughout our mountains and piedmont and less commonly in the coastal plain, Indian Tobacco is distributed throughout much of eastern and central North America south to Georgia and Texas. The common name refers to its use as a tobacco smoked for asthma and coughs. The compound lobeline is currently used, ironically, in commercial products to help people quit smoking. All parts of this plant are toxic if ingested.

JULY–FROST (178-6-8)
H-5A/SNS/RZW
[◗/N/✪/△]

Asteraceae

RAGWEED

Ambrosia artemisiifolia Linnaeus

Infamous for its power to induce allergic responses (although often attributed to the innocent Goldenrod, which blooms at the same time), this freely branched native annual is 0.2–2 m tall from a taproot. The 5–15 cm long leaves are deeply bipinnately dissected. Male and female heads are separate on the plant; the female heads are located in the axils of the upper leaves, and the male flowers, which produce the copious pollen, are in conspicuous elongate spikes.

Common on roadsides and in fields, pastures, and waste places throughout temperate North America, Ragweed is listed as an invasive weed throughout its range. The protein recognition substances found in the walls of the wind-borne pollen cause the human allergic response. Medicinal use has been made of the leaves and roots, and the pollen is used currently in the development of treatments for ragweed allergies.

AUGUST–FROST (179-2-3)
H-5A/SOL/ITG
[○/N/✪]

Asteraceae

CHICORY

Cichorium intybus Linnaeus

This introduced perennial, 3–17 dm tall, has 3–5 cm broad heads of light blue ray flowers that are borne at the ends of short spur branches and are quite conspicuous during the short time they remain open. Both basal and stem leaves are oblanceolate to lanceolate and usually deeply and irregularly lobed or cut.

This Eurasian native, and now cosmopolitan weed, is often seen on dry roadsides and in fields and waste places of North Carolina's mountains and piedmont but is absent from much of the coastal plain. Chicory has a long history of medicinal use, and the dried roots are sometimes used to flavor coffee.

MAY–FROST (179-4-1)
H-5A/SBC/HZB
[❍/D/$/✪]

Asteraceae

LION'S FOOT; GALL-OF-THE-EARTH

Prenanthes serpentaria Pursh

These interesting, freely branched native perennials with milky sap are 3–23 dm tall and have alternate, palmately lobed or dissected leaves. The drooping flower heads, grouped into a loose panicle, contain several perfect, cream-colored ray flowers, 1.5–2 cm long.

Found along roadsides and margins of deciduous woods, primarily in our mountains and piedmont, Lion's Foot ranges south to Florida and north to New Hampshire. The Cherokee used the leaves of this species for food and the roots for medicine. This is an interesting plant for the wild garden.

AUGUST–OCTOBER (179-6-6)
H-5A/SLO/PTY
[●/N/✪]

Asteraceae

WILD LETTUCE

Lactuca floridana (L.) Gaertner

These rank biennial weeds, 5–20 dm tall, have leafy stems and produce a terminal panicle up to 8 dm long and wide, which has hundreds of heads, each with 11 to 17 strap-shaped flowers that vary from blue to violet. The single-seeded fruits are plumed and wind-dispersed.

Found in mesic woodlands and open areas scattered throughout the state, Wild Lettuce ranges through the eastern and central United States north to New York and South Dakota.

AUGUST–FROST (179-7-7)
H-5A/SNL/HTB
[❍/N]

Asteraceae

SOW THISTLE

Sonchus asper (L.) Hill

The stem leaves and the larger basal rosette leaves of this introduced, glabrous winter annual have soft, spiny teeth. The stems are somewhat fleshy, usually hollow (at least near the base), and 2–3 dm tall. The flower heads of 120 or more ray flowers are 1.5–2.5 cm broad and are in branching corymbs. The swollen base, or involucre, of the flower head is quite conspicuous.

This cosmopolitan weed from Eurasia is found in lawns and waste places and along roadsides more or less throughout North Carolina. Native Americans have used Sow Thistle for both medicine and food.

APRIL–JULY (179-8-2)
H-5A/SNS/UZY
[❍/D/✪]

Asteraceae

KING DEVIL; HAWKWEED

Hieracium pratense Tausch

The leafless flowering stem of this rhizomatous perennial is 2.5–9 dm tall and bears several heads or inflorescences with strap-shaped ray flowers in a compact corymbose cluster. The basal leaves, 8–20 cm long, have bristly hairs on each surface.

An introduced European weed, ranging throughout much of northeastern North America south to Georgia, King Devil grows in pastures and clearings and along roadsides in our mountains and piedmont.

MAY–JULY (179-10-3)
H-5B/SBE/UZY
[○/N]

Asteraceae

RATTLESNAKE WEED

Hieracium venosum L.

The distinctive flat, elliptic to oblanceolate basal leaves of these herbaceous perennials are 4–15 cm long, and the veins are marked with magenta. The slender, branched flower stalk is 2–6 dm tall and bears a number of small, dandelionlike yellow heads.

Rattlesnake Weed is native to open woodlands, woodland borders, and roadsides, primarily of our mountains and piedmont and more broadly throughout the eastern United States. The leaves and roots have been used medicinally, including their use as a remedy for snakebite. If you would like to grow these interesting plants in the garden, you will need to collect seed from the wild.

APRIL–JULY (179-10-5)
H-5B/SEE/UZY
[◗/N/✪]

Asteraceae

GOAT'S BEARD

Tragopogon dubius Scopoli

These robust herbs, 5–10 dm tall, have alternate, linear leaves, 1–3 dm long, and flower heads about 2–3 cm across followed by tawny, dandelionlike fruits in a globose cluster 5–7 cm in diameter that add interest in the wild garden or a dried arrangement.

Two species of these introduced European biennials occur throughout much of the United States, but only occasionally in our area, in fields, in pastures, and along roadsides. In North Carolina, this species is known from only four counties, and the similar, but purple-flowered, *T. porrifolius* is known from only three counties.

APRIL–JULY (179-14-1)
H-5A/SLE/HZY
[◯/N]

Asteraceae

DWARF DANDELION; GOAT DANDELION

Krigia montana (Michx.) Nuttall

The leafless flowering stems of these native perennials are 1–4 dm tall, and each bears a solitary head, 2–3 cm broad, with ray flowers only. The mostly basal, linear to lanceolate leaves are 4–10 cm or more long.

These plants are native to a restricted portion of the southern Appalachians in North Carolina, Tennessee, Georgia, and South Carolina, where they grow on open rocky slopes or in the crevices of cliffs.

MAY–SEPTEMBER (179-15-2)
H-5B/SLE/HZY
[❍/N]

Asteraceae

DANDELION

Taraxacum officinale Wiggers

This naturalized, and now weedy, perennial may have been introduced into this country as an ornamental because of its bright yellow flower heads. The hollow flowering stems, 5–50 cm long, with a milky latex, have solitary, terminal heads, 2–4 cm in diameter, composed of strap-shaped ray flowers that produce a globose cluster, 5–10 cm in diameter, of wind-borne fruits. The oblanceolate and usually deeply and irregularly cut basal leaves are 5–40 cm long.

This cosmopolitan weed of temperate regions of the world, originally from Eurasia, is common in lawns, pastures, and waste places and along roadsides throughout North Carolina. Used in folk and Native American medicine for many ailments, all parts of the plant have been used as food, and the flowers are still gathered for dandelion wine. Young leaves, high in Vitamins A and C, are sometimes used as either a salad or a cooked vegetable.

FEBRUARY–JUNE (179-17-2)
H-5B/SBC/HZY
[❍/N/✪]

Asteraceae

SMALL'S RAGWORT

Packera anonyma (Wood) W. A. Weber & A. Löve
[*Senecio smallii* Britton]

These 3–8 dm tall perennials are densely pubescent at the base and have mostly elliptic-lanceolate, serrate, basal leaves, 4–15 cm long; the reduced stem leaves are deeply pinnatified. The 20 to 100 heads are arranged in terminal panicles.

Found on roadsides and in meadows, pastures, woodlands, and savannas throughout North Carolina, this Ragwort ranges through the Southeast north to Pennsylvania and Indiana. The leaves of all Ragworts of the *Packera* species have a minor liver toxicity when ingested and briefly cause skin irritation after contact.

MAY–JUNE (179-19-4)
H-5A/SES/PTY
[○/N/⚠☞]

Asteraceae

FIREWEED; BURNWEED

Erechtites hieracifolia (L.) Raf.

These tall, robust annuals, up to 3 m in height, have elliptic to lanceolate leaves, 5–20 cm long, which may be irregularly lobed. The flower heads are composed of disc flowers only, are tightly bound in an involucre (10–20 mm long) that opens, when the fruits mature, to release the dandelionlike fruits to the wind.

These plants get their common names from their appearance in disturbed or burned areas in old fields, woodlands, pastures, and waste places throughout our area. Ranging originally throughout eastern North America, Fireweed has been introduced further westward. The whole plant has been used medicinally for a variety of ailments and has potential not yet explored for modern uses.

JULY–FROST (179-20-1)
H-5A/SES/PTW
[○/N/✚]

Asteraceae

SUNBONNETS

Chaptalia tomentosa Ventenant

The leaves and stems of these 0.5–4 dm tall perennials are quite hairy, or tomentose, as indicated by the specific name. The solitary heads have both disc and ray flowers, and the 1–1.5 cm long rays usually turn deep pink on their back sides. The leaves, in basal rosettes, are elliptic to oblanceolate, distinctly white tomentose below, and 5–18 cm long.

A native of the southeastern coastal states west to Texas, Chaptalia grows on savannas and in sandy pine barrens of our coastal plain, where it reaches its northern limit. The Seminole are recorded as having used both its roots and leaves medicinally.

MARCH–MAY (179-22-1)
H-5B/SES/HZW
[◯/D/✪]

Asteraceae

CORNFLOWER; BACHELOR'S BUTTON

Centaurea cyanus L.

These naturalized, freely branched, hairy, 3–8 dm tall winter annuals have 5–15 cm long basal leaves that may be deeply dissected to unlobed; the linear, cauline leaves are unlobed. Flowers are normally blue but can vary to pink, purple, or white.

This "old timey" garden favorite, originally from Europe, reseeds readily and may spread into grasslands or pastures. Cornflower has not only escaped from cultivation throughout the United States; its seeds are often included in "wild flower mixes" for roadside planting, and it is now considered an attractive nuisance species. If you grow Cornflower in your garden, cut off and burn all flower heads before the seeds ripen!

APRIL–JUNE (179-24-2)
H-4A/SLE/HTB
[◯/N/$]

Asteraceae

BULL THISTLE

Cirsium vulgare (Savi) Ten.
[*Carduus lanceolatus* Linnaeus]

The stems and undersurface of the spiny, elliptic to lanceolate leaves of these robust, 1–2 m tall introduced biennials are covered with a thick cobweblike pubescence. The 3 to many 3–5 cm broad flower heads are in corymbs.

Found in fields, meadows, and waste places and along roadsides of our piedmont and mountains, Bull Thistle ranges throughout temperate North America. Thistles are aggressive weeds to us but are quite attractive to butterfly pollinators. This species has a long record of medicinal use as an emetic, cancer treatment, and antirheumatic by Native Americans throughout the United States. The "down" of the fruits is used by the Cherokee in their blow darts.

JUNE–FROST (179-25-1)
H-5A/SEC/HTP
[❍/N/✪]

Asteraceae

YELLOW THISTLE

Cirsium horridulum Michaux
[*Carduus spinosissimus* Walter]

These variable, robust, native annuals or biennials, 2–8 dm tall, bear 1 to 10 flower heads that can be yellow or red-purple. The spiny, elliptic leaves are villous above and have a grayish cobweblike pubescence beneath.

Found in fields, meadows, and waste places and along roadsides of our piedmont and coastal plain, they range throughout the coastal states from Maine to Texas. The Houma used the leaves and roots as an expectorant and astringent, and the Seminole used the "down" of the fruits for their blow darts.

MARCH–JUNE (179-25-3)
H-5A/SEC/HTY
[❍/N/✪]

Asteraceae

SWAMP THISTLE

Cirsium muticum Michaux
[*Carduus muticus* (Michx.) Persoon]

The heads of tubular disc flowers of this weakly spiny, 1–2.3 m tall biennial are about 2.5–3.5 cm in diameter. The hairy, elliptic to oblanceolate leaves are 1–3 dm long.

Native in much of eastern and central North America, this species of Thistle reaches its southern limit in North Carolina, where it is rare at scattered localities in bogs, meadows, and low woodland margins of our mountains and piedmont.

AUGUST–FROST (179-25-11)
H-5A/SES/HTB
[❍/M]

Asteraceae

BURDOCK

Arctium minus (Hill) Bernh.

The 5–10 dm, erect, freely branched, leafy, hollow stems of this common, introduced biennial weed arise from a basal whorl of large ovate, cordate, serrate leaves, 1.5–4.5 dm long. The spine-tipped bracts of the soft, 1.5–2.5 cm bur adhere to each other as well as to fur or hair and can form large clumps in the mane or tail of horses.

Burdock is frequent in pastures and waste places and along roadsides in North Carolina's mountains and piedmont and ranges throughout much of North America. A widely catalogued invasive alien, Burdock has a long record of medicinal use throughout its range by Native Americans: a root tea was used as a blood purifier, and poulticed leaves were used to treat rheumatism.

JUNE–FROST (179-26-3)
H-5A/SOS/HTB
[❍/N/✪]

Asteraceae

IRONWEED

Vernonia noveboracensis (L.) Michaux

These striking native perennials, 1–2 m tall, have numerous, lanceolate, cauline leaves, generally 1–2 dm long, that are often tomentose below. The 6–10 mm broad flower heads, of all disc flowers, are in a loose corymb.

An inhabitant of stream margins, meadows, pastures, and low, open woodlands throughout most of the state except the outer coastal plain, Ironweed is native to the eastern United States north to New Hampshire. Ironweed makes a spectacular display with Goldenrods and Thoroughworts in moist areas of the perennial garden, where it will be attractive to butterflies. The Cherokee used infusions of the roots to treat a variety of ailments.

JULY–SEPTEMBER (179-27-5)
H-5A/SNS/UTB
[❍/w/$/✪]

Asteraceae

ELEPHANT'S FOOT

Elephantopus tomentosus Linnaeus

Both the common and generic names of this plant refer to the large, densely tomentose, oblanceolate, basal leaves, 10–32 cm long, which do in fact resemble an elephant's foot. These native perennials, 1–6 dm tall, have only a few stem leaves, 1–7 cm long. The heads, composed entirely of disc flowers, are about 1–1.5 cm broad.

Native to the southeastern United States north to Maryland and west to Texas, these plants are common in woodlands throughout our piedmont and coastal plain.

JULY–SEPTEMBER (179-28-2)
H-5B/SBE/HTB
[◗/N]

Asteraceae

BLAZING STAR; LIATRIS

Liatris spicata (L.) Willdenow

The stiff, erect stems of this perennial, up to 2 m tall, have linear leaves, the lower leaves 1–3 dm long. The many flower heads, each 1–2 cm broad, with usually 7 to 9 disc flowers, are arranged in a dense spike.

Liatris is native to the eastern and central United States, north to Massachusetts and west to Missouri, and grows in bogs and open woodlands of our mountains and coastal plain. These plants make excellent additions to the native perennial garden, where they are quite attractive to butterflies. The Cherokee used the roots of Blazing Star as a diuretic and emetic.

AUGUST–SEPTEMBER (179-30-1)
H-5A/SNS/UTB
[❍/M/$/✪]

Asteraceae

BLAZING STAR

Liatris squarrosa (L.) Michaux

This robust, branched species of Liatris, up to 1 m tall, has narrowly elliptic leaves, the lowest to 3.5 dm long. The flower heads, 2–3 cm or more broad, are made up of 20 to 30 disc flowers. The heads are notable for the spreading or squarrose bracts beneath the flowers—thus the species name.

This perennial is native to much of the eastern and central United States and grows in North Carolina in upland woods, usually on basic soils, throughout the piedmont but is rare in the mountains and coastal plain. This is among the Liatris species recommended for the wild garden.

AUGUST–SEPTEMBER (179-30-14)
H-5A/SNS/HTB
[❍/D/$]

Asteraceae

TRILISA

Carphephorus paniculatus (J. F. Gmel.) Herbert
[*Trilisa paniculata* (J. F. Gmel.) Cassini]

These densely pubescent perennial herbs have flowering stems, 4–8 dm tall, rising from a rosette of basal leaves that are mostly 1–2 dm long. The small, often clasping, lanceolate stem leaves are closely appressed to the stem. The small heads of disc flowers only are 4–6 mm broad and are arranged in a narrow panicle.

Native to just five southeastern coastal states, these plants grow in low pinelands of our coastal plain, where they reach their northern limit. The related, glabrous species, *C. odoratissimus*, also of our coastal plain, has the scent of vanilla and is used to flavor tobacco.

SEPTEMBER–OCTOBER (179-31-2)
H-5A/SEE/HTB
[❍/w]

Asteraceae

JOE-PYE-WEED; QUEEN-OF-THE-MEADOW

Eupatorium fistulosum Barratt

These hollow-stemmed perennials, 0.5–2 m tall, have whorled, lanceolate leaves, 10–25 cm long, and a large, rounded inflorescence, up to 5 dm broad, made up of many small heads of usually 5 to 7 disc flowers.

Indigenous to the eastern United States west to Illinois and Texas, this *Eupatorium* grows in bogs, marshes, and meadows, chiefly in our mountains and piedmont. Joe-Pye-Weed can be raised from collected seed and makes a good background plant in the open, moist habitat. The closely related *E. maculatum*, with solid stems, is also a good candidate for the wild garden.

JULY–OCTOBER (179-34-3)
H-5W/SNS/HTR
[❍/w/$]

Asteraceae

DOGFENNEL; THOROUGHWORT

Eupatorium capillifolium (Lam.) Small

The one to several 1–2 m tall stems with finely dissected leaves and large terminal panicles of heads, 3–8 dm long, impart a feathery appearance that identifies this distinctive annual or weak perennial. The lobes of the creamy white flowers may be purple.

Found in fields, meadows, pastures and disturbed woods primarily in our coastal plain and piedmont, Dogfennel ranges from New Jersey throughout the Southeast to Texas. Dogfennel is a good choice for the large perennial border, and the persistent fruiting stems are attractive in the winter landscape.

SEPTEMBER–FROST (179-34-5)
H-50/SEC/HTW
[❍/N/$]

Asteraceae

BONESET

Eupatorium perfoliatum L.

The hairy, lanceolate, opposite, prominently reticulate leaves are perfoliate, or joined around the erect, solid stem, of these 6–15 dm tall perennials. The heads are in a corymb 0.6–4 dm broad.

Native in alluvial woods, bogs, and wet meadows, essentially throughout our state and much of the eastern and central United States, Boneset makes and excellent addition to the open, wild garden. Both Native Americans and early settlers treated many ailments, including flu, with the leaves of Boneset, and the immune system–enhancing properties of this species have now been verified.

AUGUST–OCTOBER (179-34-18)
H-50/SNS/HTW
[❍/W/$/✪]

Asteraceae

WHITE SNAKEROOT

Ageratina altissima (L.) King & H. E. Robins.
[*Eupatorium rugosum* Houttuyn]

This variable rhizomatous perennial, with one or several stems 3–15 dm tall, has opposite, ovate, coarsely serrate leaves mostly 6–18 cm long. Heads, each with 11 to 22 disc flowers, are in a terminal corymb.

These plants grow in rich woods and along woodland margins, chiefly in our mountains and piedmont; they are native to the northeastern United States and are good candidates for the perennial border. The root of White Snakeroot was indeed used in a poultice to treat snakebite and as a tea for several ailments. All parts of this plant are highly poisonous and may prove fatal if eaten. The milk from cows that have eaten it will cause "milk sickness" in humans.

JULY–OCTOBER (179-34-21)
H-50/SOS/HTW
[◑/M/$/✪/◉]

Asteraceae

AGERATUM; MISTFLOWER

Conoclinium coelestinum (L.) de Candolle
[*Eupatorium coelestinum* Linnaeus]

This densely pubescent, rhizomatous perennial has branched stems, 3–10 dm tall, with opposite, ovate, 4.5–9 cm long leaves. The decorative, compact, bell-shaped heads, with 40 to 50 blue disc flowers, are in a terminal corymb.

Native to the eastern and central United States north to New York and Nebraska, Ageratum grows in damp thickets and the margins of low woods of our coastal plain and, less frequently, the piedmont. Ageratum, with over a century of horticultural use, makes a colorful addition to the wild garden as well as to the perennial border.

AUGUST–OCTOBER (179-34-24)
H-50/SOS/HTB
[◗/M/$]

Asteraceae

MARSH FLEABANE

Pluchea foetida (L.) DC.

This hairy perennial, 3–9 dm tall, has thickened sessile, often clasping, elliptic to ovate leaves. The outer flowers of the heads are pistillate and filiform, giving them a feathery appearance; the inner flowers are perfect, usually sterile and tubular. The plants have a somewhat pungent, "woodsy" odor.

Pluchea grows in marshes, ditches, savannas, pocosins, and moist fields of our coastal plain and adjacent piedmont and ranges from New Jersey throughout the southeast to Texas.

JULY–OCTOBER (179-36-2)
H-5A/SED/UTW
[○/W/✪]

Asteraceae

PUSSY-TOES; EVERLASTING

Antennaria solitaria Rydberg

Identified by the solitary flower head, these low, stoloniferous perennials, 2–25 cm tall, are dioecious, producing only female flowers on one plant and only male flowers on another. The tomentose leaves are mostly 3–12 cm long.

This native of the eastern and central United States from Pennsylvania to Georgia grows in deciduous woodlands of our mountains and piedmont and, less frequently, the coastal plain. This species (and the related *A. plantaginifolia*) was once used as medicine for snakebite, lung ailments, and dysentery, among other conditions. Pussy-Toes is an ideal mat-forming plant for the wild garden.

MARCH–MAY (179-39-1)
H-5B/SBE/HTW
[◗/N/✪]

Asteraceae

RABBIT TOBACCO

Pseudognaphalium obtusifolium (L.) Hilliard & Burtt
[*Gnaphalium obtusifolium* Linnaeus]

This native annual or biennial is 3–10 dm tall. The flower heads, upper branches, and the undersurface of the narrow, 2.5–8 cm long leaves are white or grayish-white with a cottony pubescence. The pungent dried leaves account for the common name.

These variable plants grow in old fields, pastures, and waste places throughout North Carolina and eastern and central North America. Rabbit Tobacco is sometimes used in dried flower arrangements. The leaves were once chewed, or brewed as a tea, to treat a large range of ailments.

AUGUST–OCTOBER (179-40-1)
H-5A/SEE/HTW
[❍/D/✪]

Asteraceae

SEA-MYRTLE

Baccharis halimifolia Linnaeus

This 1–4 m tall shrub is one of the few woody members of the Aster family occurring in North Carolina. The elliptic to obovate leaves are 3–7 cm long and the small, white or gray heads of disc flowers are arranged in small clusters.

A somewhat weedy native of the eastern coastal states from Massachusetts to Texas, this shrub grows on roadsides and in old fields, clearings, and waste places throughout our coastal plain and is now spreading into much of the piedmont.

SEPTEMBER–OCTOBER (179-43-3)
H-5A/SES/HTW
[❍/N/$]

Asteraceae

ROBIN'S-PLANTAIN

Erigeron pulchellus Michaux

The attractive flower heads of these stoloniferous perennials have both ray and disc flowers and are generally 3–4 cm across; the villous, or hairy, stem is 1–6 dm tall.

This *Erigeron* is native to the eastern and central United States and grows in rich woods, chiefly in our mountains and piedmont. Robin's-Plantain is a good choice for the wild flower garden on well-drained soils. Another related *Erigeron* species suitable for the garden is the taller (up to 10 dm) Daisy Fleabane, *E. philadelphicus*.

APRIL–JUNE (179-44-1)
H-5B/SBS/HZW
[◗/N/$]

Asteraceae

DAISY FLEABANE

Erigeron annuus (L.) Persoon

The seeds of this "winter annual," one of several *Erigeron* species known as Daisy Fleabane, germinate and begin growth in the winter; by May or June, the first plants begin to bloom as other plants mature, carrying the blooming season to early fall. The solid stems are 0.4–1.5 m tall, villous at the base, with alternate, elliptic to oblanceolate leaves, 3–12 cm long. The flower heads have about 100 slender white pistillate flowers that vary in color from white to light lavender.

This attractive, common weed of roadsides, fields, and waste areas is found throughout our state and most of the United States and southern Canada. The seed is easy to collect if you want to try some of these plants in the garden.

MAY–OCTOBER (179-44-4)
H-5A/SES/HZW
[❍/N]

Asteraceae

WHITE WOOD ASTER

Eurybia divaricata (L.) Nesom
[*Aster divaricatus* Linnaeus]

These rhizomatous perennials, 2–10 dm tall, have glabrous stems with cordate to lanceolate leaves, up to 16 cm long, and relatively few white-flowered heads with rays up to 1.5 cm long.

Frequent in woodlands, along roadsides, and in thickets of our mountains and upper piedmont, this species ranges from Maine through the mountains to Georgia and Alabama. White Wood Aster is a good choice for the shady garden.

AUGUST–OCTOBER (179-47-6)
H-5A/SNS/HTW
[◗/N/$]

Asteraceae

NEW ENGLAND ASTER

Symphyotrichum novae-angliae (L.) Nesom
[*Aster novae-angliae* Linnaeus]

This spectacular perennial Aster with one to several erect, densely pubescent stems, 0.8–2 m tall, has up to 30 nodes below the inflorescence with 2–9 cm long, lanceolate leaves. The flower heads have 40 to 80 showy blue ray flowers that may rarely vary to pink or white.

Found in wet meadows, bogs, and marshes in our mountains, New England Aster ranges throughout much of the northern United States and southern Canada south to Georgia and also occurs in many of our western states. Widely available in the trade, this Aster is spectacular in the wild garden. Native Americans used the roots as a remedy for fevers and a variety of other ailments.

SEPTEMBER–OCTOBER (179-47-17)
H-5A/SNE/HZB
[❍/N/$/✪]

Asteraceae

WHITE-TOPPED ASTER

Symphyotrichum retroflexum (Lindl. ex DC.) Nesom
[*Aster curtisii* Torrey & Gray]

This 6–15 dm tall, profusely flowering perennial with flower heads 3–5 cm across, is one of North Carolina's most showy Asters. The linear to elliptic leaves are 6–15 dm long.

This southern Appalachian species, known from only four states, grows in woodlands, in woodland margins, and on road banks in our mountain counties, and is similar in many respects to *A. novi-belgii* of the coastal plain. These asters work well in the wild flower garden along with Goldenrods, Thoroughworts, and Sunflowers.

SEPTEMBER–OCTOBER (179-47-24)
H-5A/SES/HZB
[◗/N/$]

Asteraceae

CAROLINA ASTER; CLIMBING ASTER

Ampelaster carolinianus (Walt.) Nesom
[*Aster carolinianus* Walter]

These robust, sprawling, floriferous perennials have stems 1–2 m or more long and elliptic to lanceolate, entire leaves, 2–6 cm long. There are 50 to 70 rays per head, 1.5–2.5 cm long; the disc flowers are yellow and turn red, presumably after pollination.

State listed in North Carolina, where it is known only from low woodlands in coastal Bladen County, Climbing Aster ranges south through Florida and is an ideal choice for landscape or wild garden use—or to grow on a fence. ✗

SEPTEMBER–NOVEMBER (179-47-26)
H-5A/SEE/HZW
[○/N/$/*SR*]

Asteraceae

FROST ASTER

Symphyotrichum pilosum (Willd.) Nesom
[*Aster pilosus* Willdenow]

The multiple stems, up to 15 dm tall, of this attractive native perennial have elliptic to linear, 2–12 cm long leaves. The many heads are organized in a diffuse panicle, each head with 20 to 30 rays, 8–15 mm long. This species is variable, with a number of named varieties throughout its range.

Found in all provinces of our state in fields, meadows, and waste places, Frost Aster ranges throughout the eastern half of the United States. This Aster works well in the fall garden.

SEPTEMBER–NOVEMBER (179-47-32)
H-5A/SEE/HZW
[❍/N/$]

Asteraceae

SILVERROD; WHITE GOLDENROD

Solidago bicolor Linnaeus

The erect, 4–15 dm tall stems of these perennials have elliptic lower leaves, 6–12 cm long. The heads, with 10 to 20 flowers each, are in compact corymbs.

Silverrod grows in dry, often poor soil on road banks and in open woodlands across the northern part of all three provinces in North Carolina. It is native to much of eastern North America south to Georgia and west to Missouri. It makes a "conversation plant" in the fall wild garden.

SEPTEMBER–OCTOBER (179-49-8)
H-5A/SED/HZY
[◗/D]

Asteraceae

ROAN MOUNTAIN GOLDENROD

Solidago roanensis Porter

Named for Roan Mountain in Mitchell County, North Carolina, where it was presumably first discovered, this low, attractive Goldenrod is 3–8 dm tall and has moderately short flowering branches, giving the plant a narrow, compact, cylindrical appearance.

Native to the southern Appalachians from Pennsylvania to Georgia, these perennials grow in woodlands and along road banks only in our mountain counties. This compact Goldenrod is a great plant to try in the wild garden (or rock garden).

AUGUST–OCTOBER (179-49-10)
H-5A/SES/HZY
[○/N/$]

Asteraceae

GRAY GOLDENROD

Solidago nemoralis Aiton

These rhizomatous perennials with distinctive gray-green stems, 4–10 dm tall, clump from short rhizomes and have oval basal and cauline leaves. The lower stem leaves, 5–12 cm long, are rapidly reduced upward. The 10- to 20-flowered heads are in open panicles about as broad as long.

This widespread and somewhat variable species is native to eastern and central North America and is found throughout North Carolina, where it grows in old fields, pastures, and waste ground and on roadsides. Gray Goldenrod has potential for drier sites in the wild garden.

SEPTEMBER–OCTOBER (179-49-22)
H-5A/SES/HZY
[○/N/$/✪]

Asteraceae

ROUGH-LEAVED GOLDENROD

Solidago rugosa P. Mill.

The rugose, or rough, leaves of this rhizomatous, 3–15 dm tall perennial, account for both its common and specific names. The bright yellow flower heads organized in recurved, wandlike branches are another identifying characteristic.

Found in low woods, meadows, old fields, bogs, and pine barrens throughout North Carolina, Rough-leaved Goldenrod ranges throughout eastern North America. This Goldenrod is wonderful in the landscape and in the wild garden; a recent cultivar developed by the North Carolina Botanical Garden called "Fireworks" is readily found in the trade.

SEPTEMBER–OCTOBER (179-49-29)
H-5A/SNS/PTY
[❍/N/$/✪]

Asteraceae

TALL GOLDENROD

Solidago canadensis Linnaeus
[*Solidago altissima* Linnaeus]

These coarse, rhizomatous perennials, up to 2 m tall, have scabrous, or rough to the touch, lanceolate leaves, 6–15 cm long, with 3 prominent veins. The open, cone-shaped inflorescence, 3 dm broad and high, has branches that curve downward.

Six varieties of *S. canadensis* are found throughout North America; those in North Carolina are of the variety *scabra*, native to much of the United States but most common in the eastern deciduous forest and found throughout our state. These handsome plants make a great addition to the perennial border and are easily naturalized. Throughout the range of this species, Native Americans used its roots and flowers as an emetic and a remedy for fever, burns, snakebite, and many other ailments.

SEPTEMBER–OCTOBER (179-49-30)
H-5A/SNS/HZY
[❍/N/$/✪]

Asteraceae

SLENDER-LEAVED GOLDENROD

Euthamia tenuifolia (Pursh) Nuttall
[*Solidago tenuifolia* Pursh]

The wiry, slender stems of these 5–12 dm tall perennials branch at the top to form a more or less flat cluster of flower heads. The narrow leaves, from which the plant gets its common and specific names, are 3–8 cm long.

This native of the Atlantic Coast and Great Lakes states occurs in North Carolina only in a few of our easternmost counties, where it grows in pine barrens and brackish marshes.

SEPTEMBER–OCTOBER (179-49-37)
H-5A/SLE/HZY
[❍/N]

Asteraceae

BEARSFOOT

Smallanthus uvedalius (L.) Mackenzie ex Small
[*Polymnia uvedalia* Linnaeus]

The 1–3 m stems with large, palmately lobed, primarily opposite leaves, 1–3 dm long, and the terminal yellow heads, with rays up to 2.5 cm long, make this rank perennial weed easy to spot and identify.

Native to pastures, woodland borders, and meadows usually in low ground throughout the state, Bearsfoot ranges from New York south to Florida and Texas. If you have the space, gather some seed and grow a few, but you will have to beat the Goldfinches! The Cherokee and Iroquois used the roots and stalks of this plant to relieve back pain and rheumatism and to make a salve for burns.

JULY–OCTOBER (179-52-1)
H-5O/SOL/HZY
[❍/M/✪]

Asteraceae

GREEN-AND-GOLD

Chrysogonum virginianum Linnaeus

Early in the season the flowering stems of these pubescent perennials are only 5–10 cm long, but the stems elongate through the season, and the later-flowering stems may be 3–6 dm tall. The wide rays or "petals" of the flowering head are 1–1.5 cm long. The nutlets of this plant have an attached food body, or eliasome, that attracts ants. The ants take the seeds back to their nest, chew off the eliasome to store for food, and leave the scattered seed to germinate.

A native of the central Atlantic region from New York to Florida and west to Louisiana, *Chrysogonum* is relatively frequent in open deciduous forests of our piedmont and inner coastal plain. Green-and-Gold is an excellent choice in full or filtered sun for the wild garden, where it will form a groundcover and bloom sporadically through the season.

MARCH–JUNE (179-56-1)
H-50/SOD/HZY
[◗/N/$]

Asteraceae

WILD QUININE

Parthenium integrifolium Linnaeus

This perennial, from tuberous, thickened roots, has small white flowers borne in compact heads that are clustered in corymbs at the top of the 0.5–1.2 m tall stems. The thick, leathery leaves, 1–2.5 dm long, are scurfy or rough.

Native to the central and Mid-Atlantic states, these robust plants grow along the margins of deciduous woodlands and in old fields over much of the state except the high mountains and the southeast coastal plain. Flowering tops have been used as a treatment for intermittent fevers—thus the common name—and the roots have been used as a diuretic. There is some indication that this species has immune system–stimulating properties.

JUNE–SEPTEMBER (179-57-1)
H-5A/SNS/HZY
[○/N/$/✪]

Asteraceae

OX-EYE

Heliopsis helianthoides (L.) Sweet

As the specific name implies, this 0.3–1.5 m tall perennial looks much like some native species of *Helianthus*, or Sunflowers. The 10 or more rays or "petals," 3–5 cm long, produce a showy "flower" 8 cm or more across.

Native to eastern and central North America, this species grows in woodlands, thickets, and meadows in North Carolina's mountains and at scattered localities in our coastal plain.

JUNE–OCTOBER (179-58-1)
H-5O/SOS/HZY
[○/N]

Asteraceae

GREEN-HEADED CONEFLOWER

Rudbeckia laciniata L.

This 1–2 m tall, rhizomatous perennial has large basal leaves divided into 3 to 5 toothed lobes and smaller, alternate, ovate stem leaves. The ray flowers are yellow, but the disc flowers are green or yellowish-green, giving rise to the common name.

Found in meadows and along moist stream banks and woodland margins of our mountains and piedmont, this "green-eyed" Coneflower ranges throughout much of temperate North America. A relative of the Black-eyed Susan, it makes an attractive addition to the wild garden in shade or sun, offering a pleasing display both in leaf and flower. Native Americans used the roots, stems, and leaves to alleviate indigestion and burns, and the young leaves are still collected and cooked as greens for good health by the Cherokee, who call this plant "Souchan."

JULY–OCTOBER (179-61-1)
H-5A/SOC/HZY
[◗/M/$/✪]

Asteraceae

BLACK-EYED SUSAN; CONEFLOWER

Rudbeckia hirta Linnaeus

The round cluster of dark brown disc flowers in the center of a 12–18 cm whorl of rich golden yellow ray flowers provides the color combination that makes these 4–10 dm tall weak perennials one of our most popular, conspicuous, and well-known wild flowers.

Native over much of temperate North America (and the state flower of Maryland), these plants are rather common in our mountains and less frequent at scattered localities over the rest of the state, where they grow on roadsides and in old fields, pastures, and meadows. Two other closely related, and quite similar, later-blooming, native species, *R. fulgida* and *R. triloba*, share the common name and add to the adaptability and beauty of Black-eyed Susans for our gardens. Grow and enjoy! Native Americans used infusions of the root to wash sores and treat colds; the disc florets have been used as a dye.

JUNE–JULY (179-61-6)
H-5A/SES/HZY
[❍/N/$/✪]

Asteraceae

PURPLE CONEFLOWER; ECHINACEA

Echinacea purpurea (L.) Moench

The 5–12 dm, hirsute stems of this native perennial from thick, fleshy roots have alternate, ovate to lanceolate, 6–15 cm long leaves that are scabrous, or rough, on both surfaces. The distinctive, reflexed, 2–3 toothed, deep pink rays and darker purple disc flowers always identify this species of Coneflower.

Native in prairies and open woodlands through the central and eastern United States north to New York and Wisconsin, Purple Coneflower is rare in North Carolina, where it is found in woodlands and along road banks in only two of our mountain counties and is listed as significantly rare. Echinacea has a long history of medicinal use by Native Americans and was early incorporated into folk medicine as a virtual panacea. Now a popular herbal, it is used primarily for stimulating the immune system. Purple Coneflower is also a popular plant for the wild garden and, fortunately, is commonly propagated. Do not dig these plants from the wild. ✗

JUNE–AUGUST (179-62-1)
H-5A/SOS/HZR
[◗/N/$/✪/*SR*]

Asteraceae

NARROW-LEAVED SUNFLOWER

Helianthus angustifolius Linnaeus

The slender leaves of this tall, branched, 1–2 m tall perennial with a fibrous root system, are 6–20 cm long. The attractive heads, with rays 2–4 cm long, have reddish purple disc flowers that help distinguish this species from other Sunflowers.

A native of the southeastern states north to New York and west to Texas, this attractive and extremely floriferous Sunflower is relatively frequent in marshes, wet ditches, and low meadows throughout our coastal plain and lower piedmont but rare in the mountains. Although not well known, this species is now being used in highway plantings and is rapidly spreading along roadside drainage ditches. It is a great choice for the wild garden with moist soil.

AUGUST–OCTOBER (179-65-2)
H-5A/SLE/HZY
[○/M/$]

Asteraceae

JERUSALEM ARTICHOKE

Helianthus tuberosus Linnaeus

This coarse, hairy, rhizomatous perennial, 1–3 m tall, with the rhizomes often thickened at the ends into tubers, has lanceolate to broadly ovate leaves, 5–20 cm long, that are rough, or scabrous, above. The convex center, or disc, of the large head is 3–5 cm broad. Ray flowers are 3–5 cm long.

Native in much of temperate North America west to the Rocky Mountains, Jerusalem Artichoke is found in woodlands, clearings, and waste areas throughout the state. This sunflower is an excellent choice for the sunny border garden, mixing well with Goldenrods, Joe-Pye-Weeds, and Lobelias. The tubers have been used widely as a food and as a folk medicine for diabetes; Native Americans used the leaves, stems, and flowers to treat rheumatism.

JUNE–OCTOBER (179-65-18)
H-5A/SNS/HZY
[◗/N/$/✪]

Asteraceae

CROWNBEARD

Verbesina occidentalis (L.) Walter

The simple, glabrous, unbranched stems of these widespread perennials are 2–3 m tall with opposite, ovate or lanceolate, irregularly serrate leaves, 8–20 cm long. The yellow flower heads have perfect disc flowers and only a few, 2 to 5, sterile ray flowers.

This large weedy plant is conspicuous in large patches beginning in late summer in woodlands, fields, and pastures over much of North Carolina and ranges from Maryland to Missouri and south to Florida and Texas.

AUGUST–OCTOBER (179-66-3)
H-50/SOS/HZY
[◗/N]

Asteraceae

STAR TICKSEED; COREOPSIS

Coreopsis pubescens Elliott

These rhizomatous perennials with opposite, usually entire leaves, 4–10 cm long, are 6–10 dm tall. The conspicuously toothed rays, 3–5 cm long, and the yellow disc flowers are similar to those of most of the 15 species of Tickseed found in North Carolina. As the common name suggests, the fruits of *Coreopsis* species have 2 short, barbed awns or teeth that attach to fur, feathers, and clothing and do a great job of seed dispersal.

Native to the southeastern United States north to West Virginia, this Coreopsis grows on dry slopes and in open woods in our mountain and upper piedmont counties. As with most *Coreopsis* species, these plants make a great addition to the wild garden in drier soils.

JULY–SEPTEMBER (179-69-7)
H-50/SEE/HZY
[○/D/$]

Asteraceae

GREATER TICKSEED

Coreopsis major Walt.

The dissected, sessile, opposite leaves of this slender perennial appear whorled. The heads are corymbose and the rays, 2–4 cm long and 4–8 mm wide, are not conspicuously toothed; the disc flowers may be yellow or red.

Distributed throughout the mountains and piedmont in woodlands, thickets, and old fields in our state, Greater Tickseed ranges over much of the eastern and central United States. *Coreopsis* species are all good choices for the wild garden and *C. major* is especially suitable for drier soils.

MAY–JULY (179-69-9)
H-50/SOL/HZY
[◗/N/$]

Asteraceae

BUR MARIGOLD; TICKSEED

Bidens aristosa (Michx.) Britton

These 3–12 dm tall, profusely branched annuals with pinnately dissected leaves produce many flower heads, 5–8 cm across. The small brown fruits, each with 2 barbed teeth, stick to fur and clothing, which accounts for one of the common names.

Native through much of the eastern and central United States, *B. aristosa* is found chiefly in our piedmont and coastal plain in marshes, ditches, and meadows, where it can make spectacular displays in the fall. There are 12 species of *Bidens* in our area, including the quite weedy Spanish Needles, *B. bipinnata*, with long needlelike fruits and insignificant (rayless) flower heads.

SEPTEMBER–OCTOBER (179-70-9)
H-50/PNS/HZY
[○/M/$]

Asteraceae

BARBARA'S BUTTON; MARSHALLIA

Marshallia graminifolia (Walt.) Small

The attractive, 2–3 cm flower heads of these 4–8 dm tall perennials are made up of disc flowers only. Both the basal leaves, 5–20 cm long, and stem leaves, usually less than 3 cm long, are linear.

Indigenous in the coastal states from North Carolina to Texas, Marshallia grows in the pinelands and savannas of our coastal plain up to the Pamlico River area, where it reaches its northern limit. These plants are great possibilities for wet areas in the wild garden.

JULY–SEPTEMBER (179-72-1)
H-5A/SLE/HTR
[❍/W/$]

Asteraceae

GAILLARDIA; INDIAN BLANKET

Gaillardia pulchella Fougeroux

These hairy annuals, or short-lived perennials, 2–6 dm tall, have wide, toothed ray flowers or "petals" 1.5–2.5 cm long that are variously marked with red and yellow. The alternate, pubescent leaves are 3–8 cm long.

Native throughout much of the United States and in about a dozen counties of our coastal plain, Gaillardia grows in sandy waste places, on roadsides, and behind beach dunes. Gaillardia is an easy and colorful addition to the open wild flower garden in sandy, well-drained soils.

APRIL–OCTOBER (179-74-1)
H-5A/SNS/HZR
[❍/D/$]

Asteraceae

SNEEZEWEED

Helenium autumnale Linnaeus

The wing-angled stems of this fibrous-rooted perennial may be simple or branched above, and are 0.5–2 m tall. The numerous heads have wedge-shaped, 3-lobed rays, 1.5–3.5 cm long.

This species is native to most of the continental United States and grows in moist pastures, meadows, and ditches more or less throughout our mountains and piedmont and less frequently in the coastal plain. Along with the related *H. flexuosum*, Sneezeweeds make attractive additions to the wild garden in wet soils. The powdered leaves of this plant induce sneezing—thus the common name—and a tea made from the flowers has had medicinal use. But note that all parts of the plant are toxic when ingested due to the presence of helenalin, a lactone that is also showing potential as an antitumor agent.

SEPTEMBER–OCTOBER (179-75-1)
H-5A/SES/HZY
[◯/W/$/✪/△]

Asteraceae

YARROW; MILFOIL

Achillea millefolium Linnaeus

The numerous small flower heads of this introduced perennial, 3–12 dm tall, are each 7–11 mm across, but in the aggregate form larger clusters 5–8 cm broad. The leaves are divided into many narrow segments and have a feathery appearance.

A naturalized and somewhat weedy escape from cultivation, Yarrow grows in pastures, old fields, and waste places more or less throughout North America and all parts of our state. Yarrow is widely cultivated and a great performer in the garden, where it is tolerant of a wide range of conditions. This species has a broad spectrum of medicinal uses; over 100 active biological compounds have been identified from Yarrow. Warning: Yarrow can cause minor skin irritation and contains a known toxin, thujone.

MAY–OCTOBER (179-79-1)
H-5A/PLL/HZW
[◯/N/$/✪/△☞]

Asteraceae

DAISY; OX-EYE DAISY

Leucanthemum vulgare Lamotte [*Chrysanthemum leucanthemum* Linnaeus]

Probably one of the few flowers known to everyone, these attractive rhizomatous perennial weeds, 3–7 dm tall, have rich yellow disc flowers and bright white ray flowers or "petals" that form a flower head 5–7 cm or more across.

Like some other introductions, Daisies are thoroughly naturalized in fields, pastures, and waste places and along roadsides throughout North Carolina and most of North America. Several cultivars of Ox-eye Daisy are available for the garden.

APRIL–JULY (179-82-1)
H-5A/SBS/HZW
[❍/N/$/✪]

APPENDIX 1. HORTICULTURAL CHART FOR NATIVE NORTH CAROLINA WILD FLOWERS

Index No.[a]	Species	Bloom Time	Height[b]	Duration[c]	Light[d]	Soil[e]	Poisonous Plants[f]	Plant Status[g]	References[h] F&D	M	P	D&H
(115-1-2)	*Acer spicatum*	May–July	to 10 m	P	●	M				✔		✔
(179-79-1)	*Achillea millefolium* (I)[i]	May–Oct.	3–12 dm	P	❍	N	△☞		✔	✔	✔	
(76-4-1)	*Aconitum uncinatum*	Aug.–Sept.	to 1.5 m	P	◗	M	◉					
(76-8-1)	*Actaea pacypoda*	Apr.–May	4–8 dm	P	●	M	◉		✔	✔		
(76-9-1)	*Actaea racemosa*	May–July	1–2.5 m	P	◗	M			✔	✔		
(116-1-3)	*Aesculus flava*	Apr.–June	to 30 m	P	◗	M	◉			✔		✔
(116-1-2)	*Aesculus pavia*	Apr.–May	1–4 m	P	◗	M	◉			✔		✔
(166-25-3)	*Agalinis purpurea*	Aug.–frost	1–12 dm	A	❍	M		P				
(179-34-21)	*Ageratina altissima*	July–Oct.	3–15 dm	P	❍	M	◉		✔			
(97-9-2)	*Agrimonia parviflora*	July–Sept.	7–18 dm	P	◗	W			✔	✔		
(41-20-2)	*Aletris farinosa*	May–July	4–12 dm	P	❍	N			✔	✔		
(41-35-5)	*Allium cernuum*	July–Sept.	2–6 dm	P	❍	N			✔	✔	✔	
(41-35-7)	*Allium tricoccum*	June–July	1.5–3 dm	P	●	M			✔	✔	✔	
(179-2-3)	*Ambrosia artemisiifolia*	Aug.–frost	0.2–2 m	A	❍	N			✔	✔		
(97-21-1)	*Amelanchier arborea*	Mar.–May	to 10 m	P	◗	N				✔		✔
(41-21-1)	*Amianthium muscaetoxicum*	May–July	2.5–8 dm	P	❍	M	◉			✔		
(98-18-5)	*Amorpha fruticosa*	Apr.–June	1.5–4 m	P	◗	N				✔		✔
(179-47-26)	*Ampelaster carolinianus*	Sept.–Nov.	1–2 m	P	❍	N		SR				
(156-1-1)	*Amsonia tabernaemontana*	Apr.	4–10 dm	P	◗	M						
(29-87-7)	*Andropogon virginicus*	Sept.–Oct.	5–15 dm	P	❍	D				✔		
(76-15-1)	*Anemone acutiloba*	Mar.–Apr.	4–15 cm	P	●	M			✔			
(76-15-2)	*Anemone americana*	Feb.–Apr.	4–15 cm	P	●	M			✔			
(76-16-3)	*Anemone quinquefolia*	Mar.–May	0.5–3.5 dm	P	●	M	△					

Index No.[a]	Species	Bloom Time	Height[b]	Duration[c]	Light[d]	Soil[e]	Poisonous Plants[f]	Plant Status[g]	References[h] F&D	M	P	D&H
(76-16-1)	*Anemone virginiana*	May–July	3–10 dm	P	◗	M	△		✓	✓		
(140-32-2)	*Angelica triquinata*	Aug.–Sept.	6–20 dm	B/P	◗	M						
(179-39-1)	*Antennaria solitaria*	Mar.–May	2–25 cm	P	◗	N			✓			
(49-18-1)	*Aplectrum hyemale*	May–June	3–6 dm	P	●	M		M	✓	✓		
(156-3-3)	*Apocynum androsaemifolium*	June–Aug.	2–5 dm	P	◗	N	⦿		✓	✓		
(76-2-1)	*Aquilegia canadensis*	Mar.–May	3–12 dm	P	◗	N			✓	✓	✓	
(139-3-2)	*Aralia nudicaulis*	May–July	2–6 dm	P	●	N			✓	✓		
(139-3-3)	*Aralia racemosa*	June–Aug.	to 2.5 m	P	●	M			✓	✓		
(179-26-3)	*Arctium minus* (I)	June–frost	5–10 dm	B	❍	N			✓	✓		
(49-9-1)	*Arethusa bulbosa*	May–June	1–4 dm	P	◗	W		EM				
(32-5-1)	*Arisaema dracontium*	May	0.5–1 m	P	◗	M	△		✓	✓	✓	
(32-5-2)	*Arisaema triphyllum*	Mar.–Apr.	2–8 dm	P	●	M	△		✓	✓	✓	
(62-1-1)	*Aristolochia macrophylla*	May–June	vine	P	◗	M			✓	✓		✓
(97-14-1)	*Aruncus dioicus*	May–June	1–1.5 m	P	◗	M			✓	✓		
(62-2-1)	*Asarum canadense*	Apr.–May	1–2 dm	P	●	M			✓	✓	✓	
(157-1-11)	*Asclepias humistrata*	May–June	2–7dm	P	❍	D	△					
(157-1-1)	*Asclepias incarnata*	July–Sept.	5–15 dm	P	❍	M	△		✓	✓	✓	
(157-1-10)	*Asclepias syriaca*	June–Aug.	1–2 m	P	❍	N	△		✓	✓	✓	
(157-1-4)	*Asclepias tuberosa*	May–Aug.	2–8 dm	P	❍	D	△		✓	✓	✓	
(157-1-12)	*Asclepias variegata*	May–June	2–10 dm	P	◗	N	△					
(81-1-2)	*Asimina triloba*	Mar.–May	to 10 m	P	◗	M	△		✓	✓		✓
(94-7-1)	*Astilbe biternata*	May–July	1.5 m	P	●	M						
(98-35-2)	*Astragalus canadensis*	June–Aug.	5–15 dm	P	◗	N				✓		
(166-24-1)	*Aureolaria pedicularia*	Sept.–Oct.	3–10 dm	A	◗	D		P		✓		

Index No.[a]	Species	Bloom Time	Height[b]	Duration[c]	Light[d]	Soil[e]	Poisonous Plants[f]	Plant Status[g]	References[h]			
									F&D	M	P	D&H
(166-24-3)	*Aureolaria virginica*	May–July	3–10 dm	P	◗	D		P		✓		
(179-43-3)	*Baccharis halimifolia*	Sept.–Oct.	1–4 m	P	○	N						✓
(98-9-9)	*Baptisia alba*	May–July	5–15 dm	P	◗	D	⚠			✓	✓	
(98-9-3)	*Baptisia cinerea*	May–June	3–6 dm	P	○	D	⚠				✓	
(98-9-7)	*Baptisia tinctoria*	Apr.–Aug.	3–10 dm	P	◗	D	⚠		✓	✓		
(88-22-1)	*Barbarea verna* (I)	Mar.–June	3–8 dm	A/B	○	N				✓		
(54-3-2)	*Betula lenta*	Mar.–Apr.	to 25 m	P	●	N	⚠		✓	✓		✓
(179-70-9)	*Bidens aristosa*	Sept.–Oct.	3–12 dm	A	○	M					✓	
(167-1-1)	*Bignonia capreolata*	Apr.–May	vine	P	○	M					✓	✓
(88-8-2)	*Cakile edentula*	May–June	1–5 dm	A	○	D						
(162-4-1)	*Callicarpa americana*	June–July	1–2.5 m	P	◗	M			✓	✓		✓
(38-3-1)	*Callisia graminea*	May–July	2–5 dm	P	○	D						
(49-10-2)	*Calopogon tuberosus*	Apr.–July	3–7 dm	P	○	W		M				
(76-6-1)	*Caltha palustris*	Apr.–June	2–6 dm	P	○	W	⚠	SR	✓	✓		
(83-1-1)	*Calycanthus floridus*	Mar.–June	1–3 dm	P	◗	N	⚠		✓	✓		✓
(158-6-2)	*Calystegia sepium*	May–Aug.	vine	P	○	N			✓			
(178-2-2)	*Campanula divaricata*	July–Oct.	2–5 dm	P	◗	N				✓		
(178-2-1)	*Campanulastrum americanum*	June–Sept.	0.5–2 m	A	◗	M			✓			
(167-2-1)	*Campsis radicans*	June–July	vine	P	○	N	⚠☞				✓	✓
(88-23-7)	*Cardamine clematitis*	Apr.–May	2–4 dm	P	●	M						
(88-23-3)	*Cardamine concatenata*	Mar.–May	2–4 dm	P	●	M				✓	✓	
(179-31-2)	*Carphephorus paniculatus*	Sept.–Oct.	4–8 dm	P	○	W						
(55-2-1)	*Castanea dentata*	June–July	to 30 m	P	●	D		R		✓		✓
(55-2-2)	*Castanea pumila*	July	2–5 m	P	◗	D				✓		

Index No.[a]	Species	Bloom Time	Height[b]	Duration[c]	Light[d]	Soil[e]	Poisonous Plants[f]	Plant Status[g]	References[h] F&D	M	P	D&H
(166-28-1)	*Castilleja coccinea*	Apr.–May	2–6 dm	A/B	○	M		P	✔	✔		
(167-3-2)	*Catalpa speciosa*	May–June	to 30 m	P	○	M						✔
(77-5-1)	*Caulophyllum thalictroides*	Apr.–May	3–8 dm	P	●	M	⚠		✔	✔		
(119-2-1)	*Ceanothus americanus*	May–June	to 1 m	P	○	D			✔	✔		✔
(29-69-5)	*Cenchrus longispinus*	June–Oct.	2–5 dm	A	○	D						
(179-24-2)	*Centaurea cyanus* (I)	Apr.–June	3–8 dm	A	○	N						
(98-41-1)	*Centrosema virginianum*	June–Aug.	5–15 dm	P	◗	D						
(173-2-1)	*Cephalanthus occidentalis*	June–Aug.	1–3 m	P	○	M			✔	✔		
(98-4-1)	*Cercis canadensis*	Mar.–May	to 12 m	P	◗	N			✔	✔		✔
(98-5-5)	*Chamaecrista fasciculata*	June–Sept.	1.5–6 dm	A	○	N	⚠		✔	✔		
(41-17-1)	*Chamaelirium luteum*	Mar.–May	3–12 dm	P	●	M			✔			
(179-22-1)	*Chaptalia tomentosa*	Mar.–May	0.5–4 dm	P	○	D				✔		
(29-10-3)	*Chasmanthium latifolium*	June–Oct.	0.5–1.5 m	P	◗	M				✔		
(166-13-1)	*Chelone cuthbertii*	July–Sept.	5–10 dm	P	◗	W						
(166-13-4)	*Chelone glabra*	Aug.–Oct.	6–15 dm	P	◗	W			✔	✔	✔	
(64-3-2)	*Chenopodium album*	June–frost	to 1 m	A	○	N			✔	✔		
(145-1-1)	*Chimaphila maculata*	May–June	1–2 dm	P	●	N			✔	✔		
(153-3-1)	*Chionanthus virginicus*	Apr.–May	to 6 m	P	◗	N			✔	✔		✔
(179-56-1)	*Chrysogonum virginianum*	Mar.–June	5–10 cm	P	◗	N					✔	
(179-4-1)	*Cichorium intybus* (I)	May–frost	3–17 dm	P	○	D			✔	✔		
(179-25-3)	*Cirsium horridulum*	Mar.–June	2–8 dm	A/B	○	N				✔		
(179-25-11)	*Cirsium muticum*	Aug.–frost	1–2.3 m	B	○	M						
(179-25-1)	*Cirsium vulgare* (I)	June–frost	1–2 m	B	○	N				✔		
(98-8-1)	*Cladrastis kentukea*	Apr.–May	to 20 m	P	◗	M						

Index No.[a]	Species	Bloom Time	Height[b]	Duration[c]	Light[d]	Soil[e]	Poisonous Plants[f]	Plant Status[g]	References[h] F&D	M	P	D&H
(70-1-1)	*Claytonia virginiana*	Mar.–Apr.	0.5–4.5 dm	P	●	M				✔		
(49-8-1)	*Cleistes divaricata*	May–July	2–6 dm	P	❍	M		M				
(76-10-4)	*Clematis viorna*	May–Sept.	vine	P	◗	M	⦿			✔	✔	
(76-10-7)	*Clematis virginiana*	July–Sept.	vine	P	◗	M	⦿		✔	✔	✔	
(143-1-1)	*Clethra alnifolia*	June–July	to 2.5 m	P	❍	M						✔
(41-5-1)	*Clintonia borealis*	May–June	1.5–6 dm	P	●	M		R	✔	✔		
(41-5-2)	*Clintonia umbellulata*	May–June	2–3.5 dm	P	●	M				✔		
(98-42-1)	*Clitoria mariana*	June–Aug.	5–10 dm	P	◗	D				✔		
(107-1-1)	*Cnidoscolus stimulosus*	Mar.–Aug.	0.5–10 dm	P	❍	D	☞					
(79-1-1)	*Cocculus carolinus*	June–Aug.	vine	P	◗	N				✔		
(164-33-1)	*Collinsonia canadensis*	July–Sept.	4–15 dm	P	●	M			✔	✔		
(60-1-1)	*Comandra umbellata*	Apr.–June	1–4 dm	P	●	N		P		✔		
(38-1-3)	*Commelina communis* (I)	May–Oct.	2–8 dm	A	❍	M			✔			
(140-24-1)	*Conium maculatum* (I)	May–June	1–2 m	B	❍	N	⦿		✔	✔		
(179-34-24)	*Conoclinium coelestinum*	Aug.–Oct.	3–10 dm	P	◗	M						
(169-2-1)	*Conopholis americana*	Apr.–June	1–2 dm	P	●	D		P				
(41-11-1)	*Convallaria majuscula*	Apr.–June	1–2 dm	P	●	M	△					
(49-20-1)	*Corallorhiza maculata*	July–Aug.	1–5 dm	P	●	M		M	✔	✔		
(179-69-9)	*Coreopsis major*	May–July	0.5–1 m	P	◗	N					✔	
(179-69-7)	*Coreopsis pubescens*	July–Sept.	6–10 dm	P	❍	D					✔	
(142-1-3)	*Cornus amomum*	May–June	1–3 m	P	◗	M				✔		✔
(142-1-1)	*Cornus florida*	Mar.–Apr.	5–15 m	P	◗	N			✔	✔		✔
(86-3-1)	*Corydalis sempervirens*	Apr.–May	3–7 dm	B	❍	D				✔		
(97-20-4)	*Crataegus flabellata*	May–June	to 8 m	P	◗	N			✔			

Index No.[a]	Species	Bloom Time	Height[b]	Duration[c]	Light[d]	Soil[e]	Poisonous Plants[f]	Plant Status[g]	References[h] F&D	M	P	D&H
(107-2-2)	*Croton punctatus*	May–Nov.	to 1 m	A/P	○	D						
(158-1-7)	*Cuscuta rostrata*	Aug.–Sept.	vine	A	○	N		P				
(161-2-2)	*Cynoglosum virginianum*	Apr.–June	3–8 dm	B/P	○	N			✓	✓		
(49-1-1)	*Cypripedium acaule*	Apr.–July	2–4 dm	P	◐	M	☞	RM	✓			
(49-1-2)	*Cypripedium pubescens*	Apr.–June	2–8 dm	P	●	M	☞	RM	✓	✓		
(49-1-3)	*Cypripedium reginae*	May–June	4–10 dm	P	◐	M	☞	RM				
(111-1-1)	*Cyrilla racemiflora*	May–July	to 8 m	P	◐	M						✓
(97-6-1)	*Dalibarda repens*	June–Sept.	2–8 cm	P	◐	W		E		✓		
(165-8-1)	*Datura stramonium* (I)	July–Sept.	5–15 dm	A	○	N	⊙			✓		
(140-5-1)	*Daucus carota* (I)	May–Sept.	4–10 dm	B	○	N	☞		✓	✓	✓	
(76-3-2)	*Delphinium tricorne*	Mar.–May	2–6 dm	P	◐	M	⊙			✓		
(98-26-1)	*Desmodium nudiflorum*	July–Aug.	3–10 dm	P	◐	N			✓	✓		
(86-2-2)	*Dicentra canadensis*	Apr.–May	1–3 dm	P	●	M	△☞	R			✓	
(86-2-3)	*Dicentra cucullaria*	Mar.–Apr.	1–3 dm	P	●	M	△☞		✓	✓	✓	
(86-2-1)	*Dicentra eximia*	Apr.–June	2–5 dm	P	◐	M	△☞	SR			✓	
(174-1-2)	*Diervilla sessilifolia*	June–Aug.	0.5–2 m	P	◐	N						✓
(93-1-1)	*Dionaea muscipula*	May–June	1–3 dm	P	○	W		SR-SC		✓	✓	
(43-1-2)	*Dioscorea villosa*	Apr.–June	vine	P	●	N			✓	✓		
(77-6-1)	*Diphylleia cymosa*	May–June	4–10 dm	P	●	W		R	✓	✓	✓	
(147-1-1)	*Dodecatheon meadia*	Mar.–May	2–6 dm	P	◐	M						
(92-1-1)	*Drosera filiformis*	June	14–25 cm	P	○	M		SR			✓	
(92-1-4)	*Drosera intermedia*	July–Sept.	5–10 cm	P	○	W					✓	
(92-1-3)	*Drosera rotundifolia*	July–Sept.	10–18 cm	P	○	W		R	✓	✓	✓	
(97-2-1)	*Duchesnea indica* (I)	Feb.–frost	3–10 cm	P	○	N			✓			

Index No.[a]	Species	Bloom Time	Height[b]	Duration[c]	Light[d]	Soil[e]	Poisonous Plants[f]	Plant Status[g]	References[h]			
									F&D	M	P	D&H
(179-62-1)	*Echinacea purpurea*	June–Aug.	5–12 dm	P	◗	N		SR	✔	✔	✔	
(161-4-1)	*Echium vulgare* (I)	June–Sept.	3–6 dm	B	❍	D	⚠		✔	✔		
(179-28-2)	*Elephantopus tomentosus*	July–Sept.	1–6 dm	P	◗	N						
(169-1-1)	*Epifagus virginiana*	Sept.–Nov.	1–5 dm	A	●	M		P	✔	✔		
(145-15-1)	*Epigaea repens*	Feb.–May	2–4 dm	P	◗	D			✔	✔		✔
(137-3-2)	*Epilobium angustifolium*	July–Sept.	to 1.5 m	P	❍	N			✔	✔		
(179-20-1)	*Erechtites hieracifolia*	July–frost	to 3 m	A	❍	N			✔			
(29-85-5)	*Erianthus giganteus*	Sept.–Oct.	2–4 m	P	❍	M						
(179-44-4)	*Erigeron annuus*	May–Oct.	0.4–1.5 m	A	❍	N					✔	
(179-44-1)	*Erigeron pulchellus*	Apr.–June	1–6 dm	P	◗	N					✔	
(36-1-3)	*Eriocaulon compressum*	June–Oct.	2–8 dm	P	❍	W						
(30-9-1)	*Eriophorum virginicum*	July–Sept.	7–12 dm	P	❍	W		R				
(140-4-5)	*Eryngium integrifolium*	Aug.–Oct.	2–8 dm	P	❍	W					✔	
(140-4-3)	*Eryngium yuccifolium*	June–Aug.	2.5–12 dm	P	❍	N			✔	✔	✔	
(41-26-1)	*Erythronium americanum*	Feb.–Apr.	1–2 dm	P	●	M			✔	✔	✔	
(113-3-2)	*Euonymus americanus*	May–June	to 2 m	P	◗	N	⚠		✔	✔		✔
(179-34-5)	*Eupatorium capillifolium*	Sept.–frost	1–2 m	A/P	❍	N						
(179-34-3)	*Eupatorium fistulosum*	July–Oct.	0.5–2 m	P	❍	W						
(179-34-18)	*Eupatorium perfoliatum*	Aug.–Oct.	6–15 dm	P	❍	W			✔	✔		
(107-11-13)	*Euphorbia corollata*	May–Sept.	1–2 dm	P	❍	D	⚠		✔	✔		
(107-11-2)	*Euphorbia heterophylla*	June–Oct.	to 1 m	A	❍	D						
(107-11-10)	*Euphorbia ipecacuanhae*	Mar.–May	to 3 dm	P	❍	D			✔	✔		
(179-47-6)	*Eurybia divaricata*	Aug.–Oct.	2–10 dm	P	◗	N					✔	
(179-49-37)	*Euthamia tenuifolia*	Sept.–Oct.	5–12 dm	P	❍	N						

Index No.[a]	Species	Bloom Time	Height[b]	Duration[c]	Light[d]	Soil[e]	Poisonous Plants[f]	Plant Status[g]	References[h] F&D	M	P	D&H
(95-3-1)	*Fothergilla major*	Apr.–May	0.5–1.5 dm	P	❍	D		SR				✔
(97-1-1)	*Fragaria virginiana*	Mar.–June	2–25 cm	P	❍	N			✔	✔		
(155-3-1)	*Frasera caroliniensis*	May–June	1–3 m	P	◗	N		SR	✔	✔		
(179-74-1)	*Gaillardia pulchella*	Apr.–Oct.	2–6 dm	P	❍	D					✔	
(146-3-1)	*Galax urceolata*	May–July	2–4 dm	P	●	N				✔		
(49-2-1)	*Galearis spectabilis*	Apr.–May	1–2 dm	P	●	M		RM				
(173-10-6)	*Galium aparine*	Apr.–May	2–10 dm	A	❍	N			✔	✔		
(145-16-1)	*Gaultheria procumbens*	June–Aug.	1–2 dm	P	●	D	⦿		✔	✔		✔
(145-18-2)	*Gaylussacia dumosa*	Mar.–June	1–4 dm	P	◗	D						
(154-1-1)	*Gelsemium sempervirens*	Mar.–Apr.	vine	P	❍	M	⦿		✔	✔	✔	✔
(155-2-2)	*Gentiana quinquefolia*	Aug.–Oct.	1.5–8 dm	A	◗	M			✔	✔		
(155-2-8)	*Gentiana saponaria*	Sept.–Nov.	3–9 dm	P	◗	M		R			✔	
(155-2-4)	*Gentiana villosa*	Aug.–Nov.	1.5–5 dm	P	◗	M		R			✔	
(155-2-1)	*Gentianopsis crinita*	Sept.–Oct.	1.5–9 dm	A/B	❍	M		R				
(101-1-4)	*Geranium carolinianum*	Mar.–June	to 6 dm	A/B	❍	N						
(101-1-1)	*Geranium maculatum*	Apr.–June	3–7 dm	P	◗	M			✔	✔	✔	
(97-7-7)	*Geum radiatum*	June–Aug.	2–5 dm	P	❍	N		E-SC				
(49-13-1)	*Goodyera pubescens*	June–Aug.	2–4 dm	P	●	D		M	✔	✔		
(124-2-1)	*Gordonia lasianthus*	July–Sept.	10–15 m	P	◗	M						✔
(152-1-1)	*Halesia carolina*	Mar.–May	to 10 m	P	◗	M				✔		✔
(95-2-1)	*Hamamelis virginiana*	Oct.–Dec.	to 5 m	P	◗	N			✔	✔		✔
(179-75-1)	*Helenium autumnale*	Sept.–Oct.	0.5–2 m	P	❍	W	◬		✔	✔		
(179-65-2)	*Helianthus angustifolius*	Aug.–Oct.	1–2 m	P	❍	M					✔	
(179-65-18)	*Helianthus tuberosus*	June–Oct.	1–3 m	P	◗	N			✔	✔	✔	

Index No.[a]	Species	Bloom Time	Height[b]	Duration[c]	Light[d]	Soil[e]	Poisonous Plants[f]	Plant Status[g]	References[h] F&D	M	P	D&H
(179-58-1)	*Heliopsis helianthoides*	June–Oct.	0.3–1.5 m	P	○	N						
(41-16-1)	*Helonias bullata*	Apr.–May	1–3 dm	P	○	W		T-SC	✓			
(140-36-1)	*Heracleum lanatum*	May–July	to 2.5 m	B/P	○	N			✓	✓		
(88-17-1)	*Hesperis matronalis* (I)	May	1–1.5 m	B/P	◗	N						
(94-11-1)	*Heuchera villosa*	June–Sept.	2–7.5 dm	P	◗	N					✓	
(62-3-7)	*Hexastylis shuttleworthii*	May–July	6–10 cm	P	●	M					✓	
(122-8-2a)	*Hibiscus moscheutos*	June–Sept.	to 2 m	P	○	M			✓	✓		
(122-8-5)	*Hibiscus trionum*	June–Oct.	3–5 dm	A	○	N						
(179-10-3)	*Hieracium pratense*	May–July	2.5–9 dm	P	○	N						
(179-10-5)	*Hieracium venosum*	Apr.–July	2–6 dm	P	◗	N			✓	✓		
(173-8-1)	*Houstonia caerulea*	Apr.–May	5–10 cm	P	◗	M			✓	✓		
(173-8-5)	*Houstonia purpurea*	May–June	1–3 dm	P	◗	N						
(129-1-2)	*Hudsonia montana*	June–July	to 15 cm	P	○	D		E				
(94-5-1)	*Hydrangea arborescens*	May–July	1–3 m	P	◗	M	△		✓	✓		✓
(140-1-1)	*Hydrocotyle umbellata*	Apr.–Sept.	5–10 cm	P	○	W				✓		
(160-4-3)	*Hydrophyllum virginianum*	Apr.–June	3–8 dm	P	●	W			✓	✓	✓	
(44-3-1)	*Hymenocallis crassifolia*	May–June	3–5 dm	P	○	M	△	R				
(126-1-2)	*Hypericum hypericoides*	May–Aug.	3–10 dm	P	◗	D			✓	✓		
(126-1-8)	*Hypericum prolificum*	June–Aug.	to 2 m	P	○	M						✓
(44-7-1)	*Hypoxis hirsuta*	Mar.–June	1–4 dm	P	◗	D				✓	✓	
(112-1-6)	*Ilex montana*	Apr.–June	to 10 m	P	◗	N						
(112-1-1)	*Ilex opaca*	Apr.–June	to 15 m	P	◗	N	△		✓	✓		✓
(112-1-3)	*Ilex vomitoria*	Mar.–May	to 8 m	P	◗	M	△			✓		✓
(118-1-2)	*Impatiens capensis*	May–frost	5–15 dm	A	○	M			✓	✓		

Index No.[a]	Species	Bloom Time	Height[b]	Duration[c]	Light[d]	Soil[e]	Poisonous Plants[f]	Plant Status[g]	References[h] F&D	M	P	D&H
(118-1-1)	*Impatiens pallida*	July–Sept.	5–15 dm	A	◗	M			✓	✓		
(158-7-2)	*Ipomoea coccinea* (I)	Aug.–frost	vine	A	❍	N						
(158-7-10)	*Ipomoea macrorhiza*	June–July	vine	P	❍	D						
(158-7-11)	*Ipomoea pandurata*	May–July	vine	P	❍	D			✓	✓		
(158-7-3)	*Ipomoea purpurea* (I)	July–Sept.	vine	A	❍	N						
(46-5-8)	*Iris cristata*	Apr.–May	2–7 cm	P	◗	N	△		✓	✓	✓	
(46-5-7)	*Iris verna*	Mar.–May	2–5 cm	P	◗	D	△			✓	✓	
(46-5-2)	*Iris virginica*	Apr.–May	5–10 dm	P	◗	W	△			✓	✓	
(49-6-1)	*Isotria verticillata*	Apr.–July	1–4 dm	P	●	M		M				
(94-1-1)	*Itea virginica*	May–June	1–2 m	P	◗	M						✓
(77-4-1)	*Jeffersonia diphylla*	Mar.–Apr.	1–2 dm	P	●	M		SR	✓	✓	✓	
(145-8-2)	*Kalmia carolina*	Apr.–June	to 1.5 m	P	❍	N	⊙					
(145-8-1)	*Kalmia latifolia*	Apr.–June	2–3 m	P	◗	N	⊙		✓	✓		✓
(122-7-1)	*Kosteletskya virginica*	July–Oct.	to 1 m	P	❍	N					✓	
(179-15-2)	*Krigia montana*	May–Sept.	1–4 dm	P	❍	N						
(45-1-1)	*Lachnanthes caroliniana*	June–Sept.	3–9 dm	P	❍	W						
(179-7-7)	*Lactuca floridana*	Aug.–frost	5–20 dm	B	❍	N						
(164-19-1)	*Lamium amplexicaule* (I)	Mar.–May	1–4 dm	A/B	❍	N						
(145-7-1)	*Leiophyllum buxifolium*	Mar.–May	1–5 dm	P	◗	N				✓		✓
(33-2-2)	*Lemna perpusilla*	—	—		❍	A						
(88-4-3)	*Lepidium virginicum*	Apr.–June	1–5 dm	A	❍	N			✓	✓		
(98-27-3)	*Lespedeza procumbens*	July–Sept.	3–12 dm	P	❍	D						
(179-82-1)	*Leucanthemum vulgare* (I)	Apr.–July	3–7 dm	P	❍	N					✓	
(145-12-2)	*Leucothoe fontanesiana*	Apr.–May	to 1.5 m	P	◗	M	⊙			✓		✓

Index No.[a]	Species	Bloom Time	Height[b]	Duration[c]	Light[d]	Soil[e]	Poisonous Plants[f]	Plant Status[g]	References[h]			
									F&D	M	P	D&H
(145-12-3)	*Leucothoe recurva*	Apr.–June	to 4 m	P	◗	M						
(179-30-1)	*Liatris spicata*	Aug.–Sept.	to 2 m	P	❍	M				✔	✔	
(179-30-14)	*Liatris sqarrosa*	Aug.–Sept.	to 1 m	P	❍	D					✔	
(41-32-4)	*Lilium canadense*	June–July	6–15 dm	P	❍	M		SR	✔	✔	✔	
(41-32-2)	*Lilium catesbaei*	June–Sept.	3–10 dm	P	❍	M					✔	
(41-32-3)	*Lilium grayi*	June–July	6–12 dm	P	❍	M		T-SC			✔	
(41-32-6)	*Lilium michauxii*	July–Aug.	0.4–1.3 m	P	◗	N					✔	
(41-32-5)	*Lilium superbum*	July–Aug.	1–3 m	P	◗	N					✔	
(84-4-1)	*Lindera benzoin*	Mar.–Apr.	1–3 m	P	◗	M			✔	✔		✔
(49-15-1)	*Liparis lilifolia*	May–July	1–2 dm	P	●	N		M				
(80-1-1)	*Liriodendron tulipifera*	Apr.–June	to 60 m	P	◗	M			✔	✔		✔
(49-4-1)	*Listera smallii*	June-July	1–3 dm	P	●	M		RM				
(178-6-1)	*Lobelia cardinalis*	July–Oct.	5–15 dm	P	◗	W	◬		✔	✔	✔	
(178-6-8)	*Lobelia inflata*	July–frost	1–12 dm	A	◗	N	◬		✔	✔		
(178-6-2)	*Lobelia siphilitica*	July–Oct.	to 1 m	P	◗	M	◬		✔	✔	✔	
(174-2-4)	*Lonicera japonica* (I)	Apr.–June	vine	P	◗	N	◬		✔	✔		
(174-2-5)	*Lonicera sempervirens*	Mar.–May	vine	P	◗	N						✔
(137-1-7)	*Ludwigia leptocarpa*	June–Sept.	to 1 m	A/P	❍	W						
(98-12-1)	*Lupinus perennis*	Apr.–May	2–6 dm	P	❍	D	◬		✔	✔		
(145-11-3)	*Lyonia lucida*	Apr.–June	to 2 m	P	◗	M	⊙					
(147-2-1)	*Lysimachia ciliata*	June–Aug.	2–13 dm	P	◗	M						
(147-2-6)	*Lysimachia quadrifolia*	May–July	3–10 dm	P	❍	M			✔	✔	✔	
(80-2-4)	*Magnolia fraseri*	Apr.–May	to 15 m	P	◗	M						✔
(80-2-2)	*Magnolia grandiflora*	May–June	to 30 m	P	◗	M				✔		✔

Index No.[a]	Species	Bloom Time	Height[b]	Duration[c]	Light[d]	Soil[e]	Poisonous Plants[f]	Plant Status[g]	References[h] F&D	M	P	D&H
(80-2-1)	*Magnolia virginiana*	Apr.–June	to 20 m	P	◗	M			✔	✔		✔
(41-7-1)	*Maianthemum canadense*	May–June	4–15 cm	P	●	M			✔	✔		
(41-6-1)	*Maianthemum racemosum*	Apr.–June	2–10 dm	P	●	M			✔	✔		
(97-18-2)	*Malus angustifolia*	Apr.–May	to 10 m	P	❍	N				✔		✔
(44-8-1)	*Manfreda virginica*	May–July	1–2 m	P	◗	D			✔	✔		
(179-72-1)	*Marshallia graminifolia*	July–Sept.	4–8 dm	P	❍	W						
(157-3-5)	*Matelea carolinensis*	Apr.–June	vine	P	◗	M						
(41-4-1)	*Medeola virginiana*	Apr.–June	2–8 dm	P	●	M			✔	✔		
(41-23-2)	*Melanthium latifolium*	July–Aug.	1–2 m	P	❍	M						
(98-15-2)	*Melilotus officinalis* (I)	Apr.–Oct.	0.4–2 m	B	❍	N			✔	✔		
(177-6-1)	*Melothria pendula*	June–frost	vine	A	◗	N	⚠			✔		
(164-32-7)	*Mentha piperita* (I)	June–frost	6–16 dm	P	❍	M			✔	✔		
(145-6-1)	*Menziesia pilosa*	May–July	1–2 m	P	◗	N						
(161-10-1)	*Mertensia virginica*	Mar.–June	3–7 dm	P	◗	M		R		✔		
(98-2-1)	*Mimosa microphylla*	June–Sept.	1–2 m	P	❍	N						
(166-11-2)	*Mimulus ringens*	June–Sept.	8–10 dm	P	❍	W				✔		
(173-6-1)	*Mitchella repens*	May–June	1–3 dm	P	●	M			✔	✔		
(94-10-1)	*Mitella diphylla*	Apr.–June	2–6 dm	P	●	M				✔		
(164-23-1)	*Monarda didyma*	July–Sept.	7–18 dm	P	◗	M			✔	✔	✔	
(164-23-3)	*Monarda fistulosa*	June–Sept.	4–12 dm	P	◗	M			✔	✔	✔	
(164-23-4)	*Monarda punctata*	Aug.–Sept.	3–10 dm	P	❍	D			✔	✔		
(145-3-2)	*Monotropa hypopithys*	May–Oct.	1–3 dm	P	●	M		S		✔		
(145-3-1)	*Monotropa uniflora*	June–Oct.	1–2 dm	P	●	M		S	✔	✔		
(161-7-1)	*Myosotis scorpioides* (I)	May–Aug.	3–7 dm	P	❍	W						

Index No.[a]	Species	Bloom Time	Height[b]	Duration[c]	Light[d]	Soil[e]	Poisonous Plants[f]	Plant Status[g]	References[h] F&D	M	P	D&H
(76-12-1)	*Myosurus minimus*	Mar.–May	3–15 cm	A	❍	M				✔		
(165-2-1)	*Nicandra physalodes* (I)	July–Sept.	2–10 dm	A	❍	N						
(73-1-1)	*Nuphar advena*	Apr.–Oct.	floating	P	❍	A			✔	✔		
(166-16-1)	*Nuttallanthus texanus*	Mar.–May	1–7 dm	A/B	❍	N						
(73-2-1)	*Nymphaea odorata*	June–Sept.	floating	P	❍	A			✔	✔		
(155-5-1)	*Obolaria virginica*	Mar.–May	3–15 cm	P	●	M				✔		
(137-2-1)	*Oenothera biennis*	June–Oct.	0.5–2 m	B	❍	N			✔	✔	✔	
(137-2-9)	*Oenothera fruticosa*	Apr.–Aug.	to 1 m	P	❍	D				✔	✔	
(137-2-7)	*Oenothera speciosa*	May–Aug.	to 7.5 dm	P	❍	D						
(132-1-1)	*Opuntia humifusa*	May–June	to 4 dm	P	❍	D	☞		✔	✔	✔	
(169-3-1)	*Orobanche uniflora*	Apr.–May	7–14 cm	A	●	M		P				
(32-2-1)	*Orontium aquaticum*	Mar.–Apr.	2–6 dm	P	❍	A	⚠					
(140-8-2)	*Osmorhiza longistylis*	Apr.–May	4–8 dm	P	●	M			✔	✔		
(100-1-8)	*Oxalis grandis*	May–June	3–10 dm	A	◗	M	⚠					
(100-1-1)	*Oxalis montana*	May–Sept.	6–15 cm	P	●	M	⚠			✔		
(100-1-3)	*Oxalis violacea*	Apr.–May	1–2 dm	P	◗	N	⚠			✔		
(145-13-1)	*Oxydendrum arboreum*	June–July	to 20 m	P	◗	N			✔	✔		✔
(179-19-4)	*Packera anonyma*	May–June	3–8 dm	P	❍	N	⚠☞					
(139-2-2)	*Panax quinquefolius*	May–June	2–6 dm	P	●	M		SC	✔			
(139-2-1)	*Panax trifolius*	Apr.–June	1–2 dm	P	●	M		R	✔			
(85-5-4)	*Papaver dubium* (I)	Apr.–June	3–6 dm	A	❍	N						
(94-6-1)	*Parnassia asarifolia*	Aug.–Oct.	1–5 dm	P	❍	W		R			✔	
(179-57-1)	*Parthenium integrifolium*	June–Sept.	0.5–1.2 m	P	❍	N			✔	✔		
(120-1-1)	*Parthenocissus quinquefolia*	May–July	vine	P	◗	D	⊙		✔	✔		✔

Index No.[a]	Species	Bloom Time	Height[b]	Duration[c]	Light[d]	Soil[e]	Poisonous Plants[f]	Plant Status[g]	References[h] F&D	M	P	D&H
(131-1-1)	*Passiflora incarnata*	May–July	to 8 m	P	○	N			✓	✓	✓	
(166-1-1)	*Paulownia tomentosa* (I)	Apr.–May	to 15 m	P	○	N			✓			✓
(166-29-1)	*Pedicularis canadensis*	May–July	1–4 dm	P	◗	M			✓	✓		
(32-4-2)	*Peltandra sagittaefolia*	July–Aug.	1.5–5 dm	P	○	A						
(32-4-1)	*Peltandra virginica*	May–June	2–6 dm	P	○	A				✓		
(166-14-2)	*Penstemon canescens*	May–July	3–7 dm	P	◗	N					✓	
(160-5-1)	*Phacelia bipinnatifida*	Apr.–May	1–6 dm	B	●	M					✓	
(160-5-6)	*Phacelia fimbriata*	Apr.–May	1–5 dm	A	●	M		R			✓	
(160-5-5)	*Phacelia purshii*	May–June	1–3 dm	A	◗	M		R		✓	✓	
(94-4-2)	*Philadelphus hirsutus*	Apr.–May	1–2 m	P	○	D		R				
(159-1-9)	*Phlox carolina*	May–July	to 10 dm	P	◗	N						
(159-1-1)	*Phlox drummondii*	Apr.–July	1–7 dm	A	○	D						
(159-1-2)	*Phlox nivalis*	Mar.–May	1.5–3 dm	P	◗	D						
(97-16-1)	*Physocarpus opulifolius*	May–July	1–3 m	P	◗	M			✓	✓		✓
(164-13-1)	*Physostegia virginiana*	July–Oct.	3–10 dm	P	○	M						
(68-1-1)	*Phytolacca americana*	May–frost	1–3 m	P	○	N	⊙		✓	✓		
(145-10-1)	*Pieris floribunda*	May–June	0.2–2 m	P	○	N	⊙	R				✓
(170-1-2)	*Pinguicula lutea*	Apr.–May	1–5 dm	P	○	W				✓	✓	
(170-1-1)	*Pinguicula pumila*	Apr.–May	0.5–1.5 dm	P	○	M		SR		✓	✓	
(172-1-5)	*Plantago aristata* (I)	Apr.–July	1–2.5 dm	A	○	D				✓		
(49-3-13)	*Platanthera blephariglottis*	July–Sept.	4–8 dm	P	◗	M		RM				
(49-3-14)	*Platanthera ciliaris*	July–Sept.	2.5–10 dm	P	○	M		RM		✓		
(49-3-1)	*Platanthera lacera*	June–Aug.	2.5–7.5 dm	P	○	W		RM				
(49-3-7)	*Platanthera orbiculata*	June–Sept.	1–6 dm	P	◗	M		RM		✓		

Index No.[a]	Species	Bloom Time	Height[b]	Duration[c]	Light[d]	Soil[e]	Poisonous Plants[f]	Plant Status[g]	References[h] F&D	M	P	D&H
(49-3-4)	*Platanthera peramoena*	June–Oct.	3–10 dm	P	◗	M		RM				
(49-3-3)	*Platanthera psycodes*	June–Aug.	1.5–9.5 dm	P	◗	N		RM		✓		
(41-19-1)	*Pleea tenuifolia*	Sept.–Oct.	3–8 dm	P	○	W						
(179-36-2)	*Pluchea foetida*	July–Oct.	3–9 dm	P	○	W				✓		
(77-7-1)	*Podophyllum peltatum*	Mar.–Apr.	3–5 dm	P	◗	M	⦿		✓	✓		
(49-7-1)	*Pogonia ophioglossoides*	May–June	1–7 dm	P	○	M		M				
(106-1-7)	*Polygala curtissii*	June–Oct.	1–4 dm	A	○	N						
(106-1-14)	*Polygala lutea*	Apr.–Oct.	1–4 dm	B	○	M				✓		
(106-1-1)	*Polygala paucifolia*	Apr.–June	8–15 cm	P	◗	M				✓		
(41-10-2)	*Polygonatum biflorum*	Apr.–May	0.4–2 m	P	●	M	△		✓	✓		
(63-4-20)	*Polygonum cilinode*	June–Sept.	to 2 m	P	○	M						
(63-4-8)	*Polygonum pensylvanicum*	July–frost	0.5–1.5 m	A/P	○	M			✓	✓	✓	
(63-4-18)	*Polygonum sagittatum*	May–frost	1–2 m	A	○	M						
(39-2-1)	*Pontederia cordata*	May–Sept.	4–10 dm	P	○	A				✓		
(97-13-1)	*Porteranthus trifoliatus*	Apr.–June	4–6 dm	P	◗	M			✓	✓		
(97-4-1)	*Potentilla canadensis*	Mar.–May	to 5 dm	P	◗	N			✓	✓		
(97-4-6)	*Potentilla recta* (I)	Apr.–July	4–8 dm	P	○	D				✓		
(179-6-6)	*Prenanthes serpentaria*	Aug.–Oct.	3–23 dm	P	●	N				✓		
(41-8-2)	*Prosartes maculata*	Apr.–May	1–8 dm	P	●	M		R				
(164-12-1)	*Prunella vulgaris*	Apr.–frost	1–8 dm	P	○	N			✓	✓		
(97-22-8)	*Prunus pensylvanica*	Apr.–May	to 10 m	P	○	N	⦿			✓		✓
(97-22-11)	*Prunus serotina*	Apr.–May	to 25 m	P	◗	N	⦿		✓	✓		✓
(179-40-1)	*Pseudognaphalium obtusifolium*	Aug.–Oct.	3–10 dm	A/B	○	D			✓	✓		

Index No.[a]	Species	Bloom Time	Height[b]	Duration[c]	Light[d]	Soil[e]	Poisonous Plants[f]	Plant Status[g]	References[h]			
									F&D	M	P	D&H
(164-28-8)	*Pycnanthemum incanum*	June–Aug.	1–2 m	P	◗	N			✓	✓	✓	
(146-2-1)	*Pyxidanthera barbulata*	Mar.–Apr.	1–3 cm	P	○	D		R				
(76-13-17)	*Ranunculus bulbosus* (I)	Apr.–June	2–6 dm	P	○	N	◬			✓		
(76-13-14)	*Ranunculus recurvatus*	Apr.–June	1.5–6 dm	P	◗	M	◬			✓		
(136-1-3)	*Rhexia lutea*	Apr.–July	2–6 dm	P	○	W					✓	
(136-1-7)	*Rhexia virginica*	May–Sept.	2–9 dm	P	○	W				✓	✓	
(145-5-10)	*Rhododendron atlanticum*	Apr.–May	3–5 dm	P	◗	M	⊙					✓
(145-5-5)	*Rhododendron calendulaceum*	May–July	3–4 m	P	◗	N	⊙			✓		✓
(145-5-2)	*Rhododendron catawbiense*	Apr.–June	2–6 m	P	○	M	⊙					✓
(145-5-1)	*Rhododendron maximum*	June–July	to 10 m	P	◗	M	⊙		✓	✓		✓
(145-5-3)	*Rhododendron minus*	Apr.–June	1–3 m	P	◗	N	⊙					✓
(145-5-9)	*Rhododendron*											
	periclymenoides	Mar.–May	to 2 m	P	◗	M	⊙					✓
(145-5-4)	*Rhododendron vaseyi*	May–June	to 3 m	P	◗	N	⊙	SR				✓
(145-5-11)	*Rhododendron viscosum*	May–July	1–2 m	P	◗	D	⊙					✓
(110-1-8)	*Rhus glabra*	May–July	to 6 m	P	○	D			✓	✓		
(30-4-1)	*Rhynchospora latifolia*	May–Sept.	3–7 dm	P	○	W						
(98-32-4)	*Robinia hispida*	May–June	1–2 m	P	◗	N				✓		✓
(98-32-1)	*Robinia pseudo-acacia*	Apr.–June	to 25 m	P	○	N	◬		✓	✓		✓
(97-11-11)	*Rosa palustris*	May–July	to 2 m	P	○	M				✓		
(97-5-9)	*Rubus argutus*	Apr.–May	to 2 m	P	○	D				✓		
(97-5-1)	*Rubus odoratus*	June–Aug.	1–1.5 m	P	◗	M				✓		
(179-61-6)	*Rudbeckia hirta*	June–July	4–10 dm	P	○	N			✓	✓	✓	
(179-61-1)	*Rudbeckia laciniata*	July–Oct.	1–2 m	P	◗	M			✓	✓	✓	

Index No.[a]	Species	Bloom Time	Height[b]	Duration[c]	Light[d]	Soil[e]	Poisonous Plants[f]	Plant Status[g]	References[h] F&D	M	P	D&H
(171-3-7)	*Ruellia caroliniensis*	May–Sept.	1–6 dm	P	◗	D						
(63-2-2)	*Rumex hastatalus*	Mar.–May	4–10 dm	A	○	N						
(155-1-5)	*Sabatia angularis*	July–Aug.	3–7 dm	B	○	M				✓		
(155-1-10)	*Sabatia dodecandra*	June–Aug.	3–7 dm	P	○	W						
(27-3-8)	*Sagittaria lancifolia*	June–Oct.	6–13 dm	P	○	A			✓	✓		
(27-3-9)	*Sagittaria latifolia*	June–Sept.	6–12 dm	P	○	A				✓		
(164-22-1)	*Salvia lyrata*	Apr.–May	3–8 dm	P	○	N			✓	✓		
(174-6-1)	*Sambucus canadensis*	May–July	1–3 m	P	○	M	△		✓	✓		✓
(174-6-2)	*Sambucus pubens*	May–June	1–3 m	P	◗	M	△					✓
(85-1-1)	*Sanguinaria canadensis*	Mar.–Apr.	1–4 dm	P	●	N	⊙		✓	✓	✓	
(97-10-3)	*Sanguisorba canadensis*	July–Sept.	9–15 dm	P	○	W		R				
(71-15-1)	*Saponaria officinalis* (I)	May–frost	5–15 dm	P	○	N	△		✓	✓		
(89-1-1)	*Sarracenia flava*	Mar.–Apr.	3–10 dm	P	○	W		R			✓	
(89-1-3)	*Sarracenia minor*	Apr.–May	2–4 dm	P	○	W		SR			✓	
(89-1-4)	*Sarracenia purpurea*	Apr.–May	0.5–3 dm	P	○	W		R	✓	✓	✓	
(89-1-2)	*Sarracenia rubra*	Apr.–May	1–5 dm	P	○	W		R			✓	
(84-2-1)	*Sassafras albidum*	Mar.–Apr.	10–15 m	P	◗	N	△		✓	✓		✓
(50-1-1)	*Saururus cernuus*	May–July	5–9 dm	P	◗	A			✓	✓		
(94-14-1)	*Saxifraga michauxii*	June–Aug.	1–5 dm	P	◗	W						
(94-14-4)	*Saxifraga micranthidifolia*	May–June	3–8 dm	P	◗	W				✓		
(164-5-8)	*Scutellaria elliptica*	May–June	1.5–8 dm	P	◗	N				✓		
(91-1-9)	*Sedum telephioides*	July–Sept.	2–5 dm	P	○	D	△			✓	✓	
(91-1-5)	*Sedum ternatum*	Apr.–June	5–15 cm	P	○	M	△				✓	
(146-1-1)	*Shortia galacifolia*	Mar.–Apr.	to 18 cm	P	●	M		E-SC				

Index No.[a]	Species	Bloom Time	Height[b]	Duration[c]	Light[d]	Soil[e]	Poisonous Plants[f]	Plant Status[g]	References[h]			
									F&D	M	P	D&H
(71-17-9)	*Silene ovata*	Aug.–Sept.	3–15 dm	P	◗	M						
(71-17-1)	*Silene stellata*	July–Sept.	5–10 dm	P	◗	M				✓	✓	
(71-17-7)	*Silene virginica*	Apr.–July	2–7.5 dm	P	◗	N			✓		✓	
(46-2-4)	*Sisyrinchium angustifolium*	Mar.–June	1.5–5 dm	P	❍	N			✓	✓	✓	
(179-52-1)	*Smallanthus uvedalius*	July–Oct.	1–3 m	P	❍	M				✓		
(41-2-2)	*Smilax herbacea*	May–June	1–3 m	P	◗	M				✓		
(41-2-10)	*Smilax laurifolia*	July–Aug.	vine	P	❍	M				✓		
(165-5-5)	*Solanum carolinense*	May–July	2–8 dm	P	❍	D	⊙		✓	✓		
(179-49-8)	*Solidago bicolor*	Sept.–Oct.	4–15 dm	P	◗	D						
(179-49-30)	*Solidago canadensis*	Sept.–Oct.	to 2 m	P	❍	N				✓	✓	
(179-49-22)	*Solidago nemoralis*	Sept.–Oct.	4–10 dm	P	❍	N				✓		
(179-49-10)	*Solidago roanensis*	Aug.–Oct.	3–8 dm	P	❍	N						
(179-49-29)	*Solidago rugosa*	Sept.–Oct.	3–15 dm	P	❍	N				✓	✓	
(179-8-2)	*Sonchus asper* (I)	Apr.–July	0.3–2 m	A	❍	D				✓		
(97-19-1)	*Sorbus americana*	June–July	to 10 m	P	❍	N			✓	✓		✓
(97-19-2)	*Sorbus arbutifolia*	Mar.–May	1–2 m	P	◗	M						✓
(97-15-2)	*Spiraea tomentosa*	July–Sept.	to 2 m	P	◗	M			✓	✓		
(49-12-2)	*Spiranthes cernua*	July–frost	2–4.5 dm	P	❍	M		M	✓			
(49-12-4)	*Spiranthes lacera*	Aug.–Sept.	2–7 dm	P	◗	N		M				
(164-20-5)	*Stachys latidens*	June–Aug.	3–8 dm	P	◗	M						
(71-7-4)	*Stellaria pubera*	Apr.–June	1–4 dm	P	●	M						
(41-25-1)	*Stenanthium gramineum*	July–Sept.	3–15 dm	P	❍	M						
(124-1-1)	*Stewartia malacodendron*	May–June	to 6 m	P	◗	M						✓
(41-9-1)	*Streptopus lanceolatus*	Apr.–June	3–8 dm	P	●	M				✓		

Index No.[a]	Species	Bloom Time	Height[b]	Duration[c]	Light[d]	Soil[e]	Poisonous Plants[f]	Plant Status[g]	References[h] F&D	M	P	D&H
(98-46-2)	*Strophostyles umbellata*	June–Sept.	0.6–2 m	P	◗	N						
(98-25-1)	*Stylosanthes biflora*	June–Aug.	1–5 dm	P	◗	D				✓		
(152-2-2)	*Styrax americana*	Apr.–June	to 3 m	P	◗	M						✓
(179-47-17)	*Symphyotrichum novae-angliae*	Sept.–Oct.	0.8–2 m	P	○	N			✓	✓		
(179-47-32)	*Symphyotrichum pilosum*	Sept.–Nov.	to 15 dm	P	○	N					✓	
(179-47-24)	*Symphyotrichum retroflexum*	Sept.–Oct.	6–15 dm	P	◗	N						
(32-3-1)	*Symplocarpus foetidus*	Feb.–Mar.	4–5 dm	P	◗	W	△		✓	✓	✓	
(151-1-1)	*Symplocos tinctoria*	Mar.–May	to 10 m	P	◗	M				✓		✓
(70-2-1)	*Talinum teretifolium*	June–Sept.	1–3.5 dm	P	○	D						
(179-17-2)	*Taraxacum officinale* (I)	Feb.–June	5–50 cm	P	○	N			✓	✓		
(98-34-1)	*Tephrosia virginiana*	May–June	2–7 dm	P	○	D	△		✓	✓		
(76-11-5)	*Thalictrum revolutum*	May–July	0.5–1.5 m	P	◗	D						
(76-11-1)	*Thalictrum thalictroides*	Mar.–May	1–2 dm	P	●	M			✓	✓		
(140-15-2)	*Thaspium barbinode*	Apr.–May	5–10 dm	P	●	M				✓		
(98-10-1)	*Thermopsis villosa*	May–June	0.6–1.6 m	P	○	N					✓	
(94-12-1)	*Tiarella cordifolia*	Apr.–June	2–5 dm	P	●	M			✓	✓	✓	
(49-16-1)	*Tipularia discolor*	July–Sept.	3–5 dm	P	●	N		M				
(110-1-2)	*Toxicodendron radicans*	Apr.–May	vine	P	◗	N	☞		✓	✓		
(110-1-1)	*Toxicodendron vernix*	May–June	to 5 m	P	◗	W	☞			✓		
(38-3-2)	*Tradescantia subaspera*	June–July	3–8 dm	P	○	D	☞					
(179-14-1)	*Tragopogon dubius* (I)	Apr.–July	5–10 dm	B	○	N						
(164-1-2)	*Trichostema dichotomum*	Aug.–frost	3–7 dm	A	◗	N					✓	
(98-14-10)	*Trifolium reflexum*	Apr.–Aug.	2–5 dm	A/B	◗	N						

Index No.[a]	Species	Bloom Time	Height[b]	Duration[c]	Light[d]	Soil[e]	Poisonous Plants[f]	Plant Status[g]	References[h]			
									F&D	M	P	D&H
(41-3-7)	*Trillium catesbaei*	Apr.–June	1.5–4 dm	P	●	M					✓	
(41-3-5)	*Trillium cernuum*	Apr.–May	3–5 dm	P	●	M					✓	
(41-3-1)	*Trillium cuneatum*	Mar.–Apr.	1–3 dm	P	●	M					✓	
(41-3-3)	*Trillium discolor*	Mar.–May	1–3 dm	P	●	M		T			✓	
(41-3-6)	*Trillium erectum*	Apr.–June	2–5 dm	P	●	M				✓	✓	
(41-3-8)	*Trillium grandiflorum*	Apr.–May	2–5 dm	P	●	M				✓	✓	
(41-3-2)	*Trillium luteum*	Mar.–Apr.	1–3 dm	P	●	M		R			✓	
(41-3-10)	*Trillium undulatum*	Apr.–May	1–4.5 dm	P	●	M				✓	✓	
(41-3-6b)	*Trillium vaseyi*	Apr.–June	2–5 dm	P	●	M					✓	
(178-1-1)	*Triodanis perfoliata*	Apr.–June	1–10 dm	A	❍	N						
(49-5-1)	*Triphora trianthophora*	July–Sept.	1–3 dm	P	●	M		RM				
(19-1-1)	*Typha latifolia*	May–July	1–2.5 m	P	❍	A			✓	✓		
(29-10-4)	*Uniola paniculata*	June–July	1–2 m	P	❍	D						
(59-2-1)	*Urtica dioica* (I)	May–July	1–1.3 m	P	◗	N	☞		✓	✓		
(170-2-5)	*Utricularia inflata*	May–Nov.	4–20 cm	P	❍	A						
(170-2-4)	*Utricularia purpurea*	May–Sept.	3–15 cm	A/P	❍	A						
(41-33-2)	*Uvularia grandiflora*	Apr.–May	2–7.5 dm	P	●	M			✓	✓		
(145-19-8)	*Vaccinium corymbosum*	Feb.–May	1–4 m	P	◗	M				✓		✓
(145-19-13)	*Vaccinium erythrocarpum*	May–July	3–15 dm	P	●	M						
(145-19-14)	*Vaccinium macrocarpon*	May–July	to 2 dm	P	❍	W		SR		✓		✓
(145-19-2)	*Vaccinium stamineum*	Apr.–June	1–5 m	P	◗	D						
(41-24-1)	*Veratrum viride*	June–Aug.	6–15 dm	P	●	M	⊙		✓	✓		
(166-12-3)	*Verbascum thapsus* (I)	June–Sept.	1–2 m	B	❍	D			✓	✓	✓	
(162-1-1)	*Verbena brasiliensis* (I)	May–Oct.	1–2.5 m	P	❍	D						

Index No.[a]	Species	Bloom Time	Height[b]	Duration[c]	Light[d]	Soil[e]	Poisonous Plants[f]	Plant Status[g]	References[h] F&D	M	P	D&H
(179-66-3)	*Verbesina occidentalis*	Aug.–Oct.	2–3 m	P	◗	N						
(179-27-5)	*Vernonia noveboracensis*	July–Sept.	1–2 m	P	❍	W				✓	✓	
(166-20-5)	*Veronica persica* (I)	Mar.–June	1–4 dm	A	❍	N						
(174-5-9)	*Viburnum acerifolium*	Apr.–June	1–2 m	P	◗	M				✓		✓
(174-5-1)	*Viburnum lantanoides*	Apr.–June	to 5 m	P	◗	M						✓
(174-5-3)	*Viburnum nudum*	Apr.–May	to 4 m	P	◗	N			✓	✓		✓
(98-36-12)	*Vicia caroliniana*	Apr.–June	3–10 dm	P	◗	N				✓		
(130-2-28)	*Viola bicolor*	Mar.–May	0.5–4 dm	A	❍	N						
(130-2-21)	*Viola canadensis*	Apr.–July	1–5 dm	P	●	M			✓	✓		
(130-2-19)	*Viola hastata*	Mar.–May	0.5–2.5 dm	P	◗	M						
(130-2-16)	*Viola lanceolata*	Mar.–May	5–15 cm	P	❍	M						
(130-2-1)	*Viola pedata*	Mar.–May	5–10 cm	P	◗	D				✓	✓	
(130-2-23)	*Viola rostrata*	Apr.–May	5–12 cm	P	●	M		R				
(130-2-3)	*Viola sororia*	Feb.–May	5–15 cm	P	◗	N					✓	
(130-2-6)	*Viola villosa*	Feb.–Apr.	5–10 cm	P	❍	M		R				
(120-2-1)	*Vitis rotundifolia*	May–June	vine	P	◗	N				✓		
(98-31-1)	*Wisteria frutescens*	Apr.–May	2–15 m	P	◗	N	△					✓
(33-3-1)	*Wolffia columbiana*	--	--		❍	A						
(76-1-1)	*Xanthorhiza simplicissima*	Apr.–May	3–5 dm	P	●	M			✓	✓		✓
(41-15-1)	*Xerophyllum asphodeloides*	May–June	0.8–1.5 m	P	❍	D		R				
(35-1-1)	*Xyris fimbriata*	Sept.–Oct.	5–12 dm	P	❍	W						
(41-12-3)	*Yucca filamentosa*	May–June	1–3 m	P	❍	D			✓	✓		✓
(145-9-1)	*Zenobia pulverulenta*	Apr.–June	to 2 m	P	◗	M						✓
(44-4-1)	*Zephyranthes atamasco*	Mar.–Apr.	1–2.5 dm	P	❍	M	⦿				✓	

Index No.[a]	Species	Bloom Time	Height[b]	Duration[c]	Light[d]	Soil[e]	Poisonous Plants[f]	Plant Status[g]	References[h]			
									F&D	M	P	D&H
(41-22-4)	*Zigadenus leimanthoides*	July–Aug.	4–7.5 dm	P	❍	M	◉	T-SC				
(29-65-1)	*Zizania aquatica*	May	2–3 m	A	❍	A				✔		
(140-14-2)	*Zizia aurea*	Apr.–May	to 8 dm	P	◗	M	⚠		✔	✔		

a. Index number as shown in Radford, Ahles, and Bell, *Manual of the Vascular Flora of the Carolinas*.

b. Height is the general vertical distance from the ground to the top of the upright stems or leaves. Because of the variable and usually indeterminate length to which vines may grow, stem length is not given here for vines.

c. Duration: A = annual; B = biennial; P = perennial.

d. Light Requirements: ❍ = full sun; ● = full shade; ◗ = partial shade and sun or filtered sun all day.

e. Soil Moisture: A = aquatic; W = habitat wet most of the year; M = moist, soil should not dry out; N = average, or variable soil moisture; D = dry soils.

f. Poisonous Plants: ◉ = poisonous when ingested; ⚠ = poisonous when ingested but either of low toxicity or requires eating large quantities to achieve toxicity; ☞ = toxicity is external, causing a skin reaction.

g. Plant Status: E = endangered (species in jeopardy); T = threatened (species likely to become endangered in the foreseeable future); SC = endangered and threatened species that may be propagated under permit; SR = significantly rare; R = rare in North Carolina based on the authors' field experience; P = parasitic; M = mutualistic relationship with fungal partners (mycorrhizae); S = saprophytic.

h. References for medicinal uses and propagation: F&D = Foster and Duke, *A Field Guide to Medicinal Plants and Herbs*; M = Moerman, *Native American Ethnobotany*; P = Phillips, *Growing and Propagating Wild Flowers*; D&H = Dirr and Heuser, *The Reference Manual of Woody Plant Propagation* (see References in the Introduction for complete bibliographical information on all of these sources).

i. (I) = introduced plants now naturalized in North Carolina.

APPENDIX 2.
ENDANGERED AND THREATENED NORTH CAROLINA WILD FLOWERS

Species	Common Name	State Status[a]	Federal Status[b]	State Rank[c]	Global Rank[d]
Ampelaster carolinianus	Carolina Aster	SR-P		SH	G5
Arethusa bulbosa	Bog Rose	E		S1	G4
Caltha palustris	Marsh Marigold	SR-P		S1	G5
Dalibarda repens	Robin Runaway	E		S1	G5
Dicentra eximia	Bleeding Heart	SR-P		S2	G4
Dionaea muscipula	Venus Flytrap	SR-L-SC	FSC	S3	G3
Drosera filiformis	Threadleaf Sundew	SR-P		S1	G4/G5
Echinacea purpurea	Purple Coneflower	SR-P		S1	G4
Fothergilla major	Witch Alder	SR-T		S2	G3
Frasera caroliniensis	Columbo	SR-P		S2/S3	G5
Gentianopsis crinita	Fringed Gentian	E-SC		S1	G5
Geum radiatum	Spreading Avens	E-SC	E	S1	G1
Helonias bullata	Swamp Pink	T-SC	T	S2	G3
Hudsonia montana	Mountain Golden Heather	E	T	S1	G1
Jeffersonia diphylla	Twinleaf	SR-P		S1	G5
Lilium canadense	Canada Lily	SR-P		S1	G5
Lilium grayi	Gray's Lily	T-SC	FSC	S3	G3
Panax quinquefolius	Ginseng	SC			G3/G4
Parnassia caroliniana	Carolina Grass-of-Parnassus	E		S2	G3
Parnassia grandifolia	Large-leaved Grass-of-Parnassus	E	FSC	S2	G3
Pinguicula pumila	Butterwort	SR-P		S2	G4
Pyxidanthera brevifolia	Sandhills Pixie Moss	E	FSC	S2	G2
Rhododendron vaseyi	Pinkshell Azalea	SR-L		S3	G3
Sarracenia minor	Hooded Pitcher Plant	SR-P		S1	G4
Shortia galacifolia	Oconee Bells	E-SC	FSC	S1	G2
Trillium discolor	Mottled Trillium	T		S1	G3
Vaccinium macrocarpon	Cranberry	SR-P		S2	G4
Zigadenus leimanthoides	Deathcamus	SR		S1	G4

Sources: North Carolina Natural Heritage Program online database (www.ils.unc.edu/parkproject.nhp/); North Carolina Department of Agriculture and Consumer Services Plant Conservation Program (www.ncagr.com/plantind/plant/conserv/cons.htm).

a. State Status: E = endangered; T = threatened; SC = Special Concern; SR = Significantly Rare; L = limited range; P = rare at periphery.

b. Federal Status: E = endangered; T = threatened; FSC = federal species of concern (species at risk).
c. State Rank: S1 = critically imperiled; S2 = imperiled; S3 = rare or uncommon; SH = of historical occurrence (suspected to be extant).
d. Global Rank: G1 = critically imperiled; G2 = imperiled; G3 = very rare and local; G4 = globally secure but rare in restricted areas; G5 = demonstrably secure globally, may be rare at the periphery.

APPENDIX 3.
NOMENCLATURAL CHANGES FOR NORTH CAROLINA WILD FLOWERS

Index No.	Current Name	Former Name
(76-9-1)	*Actaea racemosa* Linnaeus	*Cimicifuga racemosa* Nuttall
(116-1-3)	*Aesculus flava* Aiton	*Aesculus octandra* Marshall
(179-34-21)	*Ageratina altissima* (L.) King & H. E. Robins.	*Eupatorium rugosum* Houttuyn
(179-47-26)	*Ampelaster carolinianus* (Walt.) Nesom	*Aster carolinianus* Walter
(76-15-1)	*Anemone acutiloba* (DeCandolle) G. Lawson	*Hepatica acutiloba* DeCandolle
(76-15-2)	*Anemone americana* (DeCandolle) H. Hara	*Hepatica americana* (DC) Ker
(167-1-1)	*Bignonia capreolata* Linnaeus	*Anisostichus capreolata* (L.) Bureau
(38-3-1)	*Callisia graminea* (Small) G. Tucker	*Tradescantia rosa var. graminea* (Small) Anderson & Woodson
(49-10-2)	*Calopogon tuberosus* (L.) B.S.P.	*Calopogon pulchellus* (Salisb.) R. Brown
(178-2-1)	*Campanulastrum americanum* (L.) Small	*Campanula americana* Linnaeus
(88-23-7)	*Cardamine concatenata* (Michx.) O. Schwarz	*Dentaria laciniata* Willdenow
(179-31-2)	*Carphephorus paniculatus* (J. F. Gmel.) Herbert	*Trilisa paniculata* (J. F. Gmel.) Cassini
(98-5-5)	*Chamaecrista fasciculata* (Michx.) Greene	*Cassia fasciculata* Michaux
(29-10-3)	*Chasmanthium latifolium* (Michx.) Yates	*Uniola latifolia* Michaux
(179-25-3)	*Cirsium horridulum* Michaux	*Carduus spinosissimus Walter*
(179-25-11)	*Cirsium muticum* Michaux	*Carduus muticus* (Michx.) Persoon
(179-25-1)	*Cirsium vulgare* (Savi) Ten.	*Carduus lanceolatus* Linnaeus
(98-8-1)	*Cladrastis kentukea* (Dum.-Cours.) Rudd	*Cladrastis lutea* (Michx. f.) K. Koch
(179-34-24)	*Conoclinium coelestinum* (L.) de Candolle	*Eupatorium coelestinum* Linnaeus
(41-11-1)	*Convallaria majuscula* Greene	*Convallaria montana* Rafinesque
(49-1-2)	*Cypripedium pubescens* Willdenow	*Cypripedium calceolus* var. *pubescens* (Willd.) Correll
(179-47-6)	*Eurybia divaricata* (L.) Nesom	*Aster divaricatus* Linnaeus
(179-49-37)	*Euthamia tenuifolia* (Pursh) Nuttall	*Solidago tenuifolia* Pursh
(155-3-1)	*Frasera caroliniensis* Walter	*Swertia caroliniensis* (Walter) Kuntze
(146-3-1)	*Galax urceolata* (Poir.) Brummitt	*Galax aphylla* Linnaeus
(49-2-1)	*Galearis spectabilis* (L.) Rafinesque	*Orchis spectabilis* Linnaeus
(155-2-1)	*Gentianopsis crinita* (Froel.) Ma	*Gentiana crinita* Froelich
(140-36-1)	*Heracleum lanatum* Michaux	*Heracleum maximum* Bartram
(145-8-2)	*Kalmia carolina* Small	*Kalmia angustifolia* var. *caroliniana* (Small) Fernald
(179-82-1)	*Leucanthemum vulgare* Lamotte	*Chrysanthemum leucanthemum* Linnaeus

Index No.	Current Name	Former Name
(145-12-2)	*Leucothoe fontanesiana* (Steud.) Sleumer	*Leucothoe axillaris* (Llam.) D. Don
(41-6-1)	*Maianthemum racemosum* (L.) Link	*Smilicina racemosa* (L.) Desfontaines
(44-8-1)	*Manfreda virginica* (L.) Salisbury ex Rose	*Agave virginica* Linnaeus
(41-23-2)	*Melanthium latifolium* Desrousseaux	*Melanthium hybridum* Walter
(98-2-1)	*Mimosa microphylla* Dryander	*Schrankia microphylla* (Smith) Macbride
(73-1-1)	*Nuphar advena* (Aiton) W. T. Aiton	*Nuphar luteum* (L.) Sibthorp & Smith
(166-16-1)	*Nuttallanthus texanus* (Scheele) D. A. Sutton	*Linaria canadensis* var. *texana* (Scheele) Pennell
(132-1-1)	*Opuntia humifusa* (Raf.) Rafinesque	*Opuntia compressa* (Salisb.) Macbride
(100-1-1)	*Oxalis montana* Rafinesque	*Oxalis acetosella* Linnaeus
(179-10-4)	*Packera anonyma* (Wood) W. A. Weber & A. Löve	*Senecio smallii* Britton
(164-13-1)	*Physostegia virginiana* (L.) Bentham	*Dracocephalum virginianum* Linnaeus
(49-3-13)	*Platanthera blephariglottis* (Willd.) Lindley	*Habenaria blephariglottis* (Willd.) Hooker
(49-3-14)	*Platanthera ciliaris* (L.) Lindley	*Habenaria ciliaris* (L.) R. Brown
(49-3-1)	*Platanthera lacera* (Michx.) G. Don	*Habenaria lacera* (Michx.) Loddiges
(49-3-7)	*Platanthera orbiculata* (Pursh) Lindley	*Habenaria orbiculata* (Pursh) Torrey
(49-3-4)	*Platanthera peramoena* (Gray) Gray	*Habenaria peramoena* Gray
(49-3-3)	*Platanthera psycodes* (L.) Lindley	*Habenaria psycodes* (L.) Sprengel
(97-13-1)	*Porteranthus trifoliatus* (L.) Britton	*Gillenia trifoliata* (L.) Moench
(41-8-2)	*Prosartes maculata* (Buckl.) A. Gray	*Disporum maculatum* (Buckl.) Britton
(179-40-1)	*Pseudognaphalium obtusifolium* (L.) Hilliard & Burtt	*Gnaphalium obtusifolium* Linnaeus
(145-5-8)	*Rhododendron periclymenoides* (Michx.) Shinners	*Rhododendron nudiflorum* (L.) Torrey
(30-1-5)	*Rhynchospora latifolia* (Baldw. ex Ell.) Thomas	*Dichromena latifolia* Baldwin
(27-3-8)	*Sagittaria lancifolia* Linnaeus	*Sagittaria falcata* Pursh
(179-52-1)	*Smallanthus uvedalius* (L.) Mackenzie ex Small	*Polymnia uvedalia* Linnaeus
(179-49-30)	*Solidago canadensis* Linnaeus	*Solidago altissima* Linnaeus
(49-12-4)	*Spiranthes lacera* (Raf.) Rafinesque	*Spiranthes gracilis* (Bigel.) Beck
(41-9-1)	*Streptopus lanceolatus* var. *roseus* (Michx.) Reveal	*Streptopus roseus* Michaux
(155-3-1)	*Swertia caroliniensis* (Walter) Kuntze	*Frasera caroliniensis* Walter

Index No.	Current Name	Former Name
(179-47-17)	*Symphyotrichum novae-angliae* (L.) Nesom	*Aster novae-angliae* Linnaeus
(179-47-32)	*Symphyotrichum pilosum* (Willd.) Nesom	*Aster pilosus* Willdenow
(179-47-24)	*Symphyotrichum retroflexum* (Lindl. ex DC.) Nesom	*Aster curtisii* Torrey & Gray
(110-1-2)	*Toxicodendron radicans* (L.) Kuntze	*Rhus radicans* Linnaeus
(110-1-1)	*Toxicodendron vernix* (L.) Kuntze	*Rhus vernix* Linnaeus
(41-3-2)	*Trillium luteum* (Muhl.) Harbison	*Trillium viride* var. *luteum* Beck
(178-1-1)	*Triodanis perfoliata* (L.) Nieuwand	*Specularia perfoliata* (L.) A. DeCandolle
(174-5-1)	*Viburnum lantanoides* Michaux	*Viburnum alnifolium* Marshall
(130-2-28)	*Viola bicolor* Pursh	*Viola rafinesquii* Greene
(130-2-3)	*Viola sororia* Willdenow	*Viola papilionacea* Pursh

INDEX

Note: Brackets around a scientific name indicate that the name, in use at the time of the first edition of *Wild Flowers of North Carolina*, has been superseded by a new scientific name.

PHOTO CREDITS

All photographs by William S. Justice except as noted below.

C. Ritchie Bell: *Peltandra sagittaefolia* (32-4-2); *Arisaema triphyllum* (32-5-2), fruit; *Commelina communis* (38-1-3); *Callisia graminea* (38-3-1); *Pontederia cordata* (39-2-1); *Trillium undulatum* (41-3-10); *Medeola virginiana* (41-4-1); *Maianthemum racemosum* (41-6-1); *Erythronium americanum* (41-26-1); *Lilium grayi* (41-32-3); *Lilium superbum* (41-32-5); *Uvularia grandiflora* (41-33-2); *Hymenocallis crassifolia* (44-3-1); *Manfreda virginica* (44-8-1); *Sisyrinchium angustifolium* (46-2-4); *Platanthera orbiculata* (49-3-7); *Platanthera ciliaris* (49-3-14); *Tipularia discolor* (49-16-1), flower; *Betula lenta* (54-3-2); *Aristolochia macrophylla* (62-1-1); *Hexastylis shuttlesworthii* (62-3-7); *Claytonia virginica* (70-1-1); *Stellaria pubera* (71-7-4); *Nuphar advena* (73-1-1); *Nympaea odorata* (73-2-1); *Aquilegia canadensis* (76-2-1); *Actaea pacypoda* (76-8-1), fruit; *Ranunculus recurvatus* (76-13-14); *Ranunculus bulbosus* (76-13-17), field; *Caulophyllum thalictroides* (77-5-1), flower; *Diphylleia cymosa* (77-6-1); *Magnolia virginiana* (80-2-1); *Magnolia fraseri* (80-2-4); *Calycanthus floridus* (83-1-1); *Sassafras albidum* (84-2-1), fruit, foliage; *Lindera benzoin* (84-4-1), fruit; *Sanguinaria canadensis* (85-1-1), leaf; *Hesperis matronalis* (88-17-1); *Sarracenia flava* (89-1-1), leaves; *Sarracenia rubra* (89-1-2); *Sarracenia minor* (89-1-3); *Itea virginica* (94-1-1); *Mitella diphylla* (94-10-1); *Hamamelis virginiana* (95-2-1); *Fragaria virginiana* (91-1-1); *Porteranthus trifoliatus* (97-13-1); *Aruncus dioicus* (97-14-1); *Spiraea tomentosa* (97-15-2); *Cercis canadensis* (98-4-1), habit; *Chamaecrista fasciculata* (98-5-5); *Baptisia alba* (98-9-9); *Amorpha fructicosa* (98-18-5); *Stylosanthes biflora* (98-25-1); *Strophostyles umbellata* (98-46-2); *Polygala paucifolia* (106-1-1); *Polygala lutea* (106-1-14); *Euphorbia heterophylla* (107-11-2); *Toxicodendron radicans* (110-1-2); *Cyrilla racemiflora* (111-1-1); *Ilex opaca* (112-1-1), fruit; *Ilex vomitoria* (112-1-3); *Euonymus americanus* (113-3-2), fruit; *Aesculus pavia* (116-1-2); *Impatiens pallida* (118-1-1); *Impatiens capensis* (118-1-2); *Vitis rotundifolia* (120-2-1); *Kosteletskya virginica* (122-7-1); *Aralia nudicaulis* (139-3-2); *Eryngium integrifolium* (140-4-5); *Angelica triquinata* (140-32-2), close-up; *Heracleum lanatum* (140-36-1); *Monotropa hypopithys* (145-3-2); *Zenobia pulverulenta* (145-9-1); *Pieris floribunda* (145-10-1); *Leucothoe recurva* (145-12-3); *Oxydendrum arboreum* (145-13-1), fruit; *Gaultheria procumbens* (145-16-1), fruit; *Vaccinium stamineum* (145-19-2); *Vaccinium erythrocarpum* (145-19-13), fruit; *Symplocos tinctoria* (151-1-1); *Sabatia dodecandra* (155-1-10); *Frasera caroliniensis* (155-3-1); *Amsonia tabernaemontana* (156-1-1); *Asclepias incarnata* (157-1-1), fruit; *Asclepias tuberosa* (157-1-4); *Asclepias syriaca* (157-1-10), fruit; *Asclepias variegata* (157-1-12); *Calystegia sepium* (158-6-2); *Ipomoea coccinea* (158-7-2); *Ipomoea purpurea* (158-7-3); *Phacelia bipinnatifida* (160-5-1); *Phacelia purshii* (160-5-5); *Phacelia fimbriata* (160-5-6); *Echium vulgare* (161-4-1); *Lamium amplexicaule* (164-19-1); *Monarda fistulosa* (164-23-3); *Pycnanthemum incanum* (164-28-8); *Aureolaria virginica* (166-24-3); *Pedicularis canadensis* (166-29-1); *Bignonia capreolata* (167-1-1); *Campsis radicans* (167-2-1); *Ruellia caroliniensis* (171-3-7); *Houstonia caerulea* (173-8-1); *Lonicera sempervirens* (174-2-5); *Viburnum acerifolium* (174-5-9); *Lobelia cardinalis*

(178-6-1); *Prenanthes serpentaria* (179-6-6); *Sonchus asper* (179-8-2); *Centaurea cyanus* (179-24-2); *Cirsium vulgare* (179-25-1); *Arctium minus* (179-26-3); *Veronia noveboracensis* (179-27-5); *Liatris squarrosa* (179-30-14); *Eupatorium fistulosum* (179-34-3); *Ageratina altissima* (179-34-21); *Solidago roanensis* (179-49-10); *Solidago rugosa* (179-49-29); *Solidago canadensis* (179-49-30); *Chrysogonum virginianum* (179-56-1); *Rudbeckia hirta* (179-61-6); *Coreopsis major* (179-69-9); *Gaillardia pulchella* (179-74-1); *Leucanthemum vulgare* (179-82-1).

Anne H. Lindsey: *Typha latifolia* (19-1-1); *Chasmanthium latifolium* (29-10-3); *Tipularia discolor* (49-16-1), leaf; *Chenopodium album* (64-3-2); *Zizia aurea* (140-14-2); *Elephantopus tomentosus* (179-28-2); *Ampelaster carolinianus* (179-47-26).

General Key Characters		
Habit	**Petal Number**	**Leaf Arrangement**
T tree	N none, or not evident	N none, or not evident
S shrub	3	O opposite
V vine	4	A alternate
H terrestrial herb	5	W whorled
A aquatic herb	6	B basal
E epiphyte, parasite	V 7+ or variable	

Key Leaf Characters	
Leaf Type	**Leaf Shape**
S simple	G needle, scale, or grass-like
P pinnate	L linear
B bipinnate	N lanceolate
M palmate	E elliptic
T trifoliate	O ovate
	C hastate, cordate
	R round
	B obovate

10 centimeters = 1 decimeter

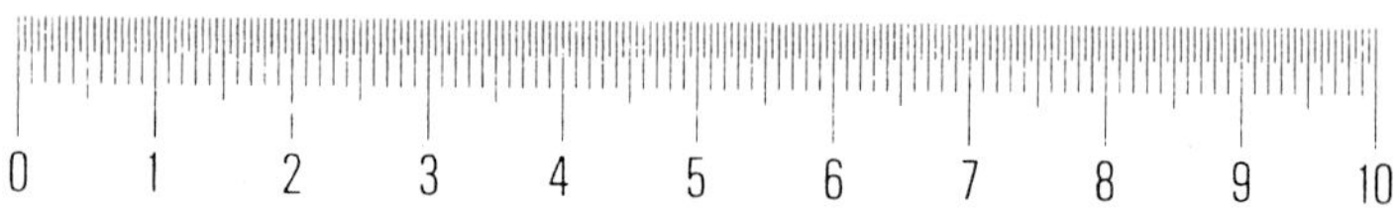

0 1 2 3 4 5 6 7 8 9 10

Leaf Margin
 E entire
 S serrate
 D dentate
 C cut
 L lobed

Key Flower Characters		
Flower Arrangement	Flower Form	Flower Color
S  solitary	N none, or not evident	G green
I  spike	R  round flat	W white, cream
R 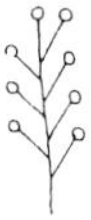 raceme	C  cup	Y yellow, orange
P 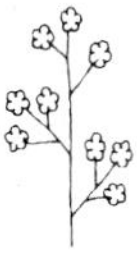 panicle	F 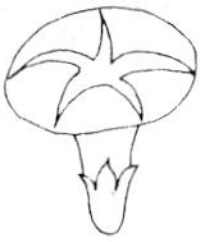 round funnel	R pink, red
U 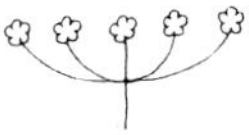 umbel, corymb, cyme	T  round tube	B blue, purple
H  head & rays	Z 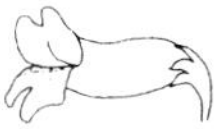 two lipped, zygomorphic	M maroon, brown
K  knot, glomerule, head		
T 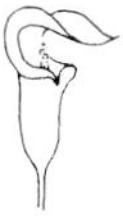 spathe